Disaster Study Number 13
Disaster Research Group
Division of Anthropology and Psychology

INDIVIDUAL AND GROUP BEHAVIOR IN A
COAL MINE DISASTER

Editors

H D BEACH
Associate Professor of Psychology
Dalhousie University
Nova Scotia

and

R A LUCAS
Associate Professor of Sociology
Acadia University
Nova Scotia

Publication 834
National Academy of Sciences—National Research Council
Washington, D. C.
1960

PROJECT STAFF

(Mrs.) Nellen Armstrong, Assistant in Sociology, Acadia University, Wolfville, Nova Scotia

H. D. Beach, Associate Professor of Psychology, Dalhousie University, Halifax, Nova Scotia

L. R. Denton, Clinical Psychologist, Nova Scotia Department of Welfare, Truro, Nova Scotia

R. A. Lucas, Associate Professor of Sociology, Acadia University, Wolfville, Nova Scotia

P. N. Murphy, Psychiatric Resident, Dalhousie University, Halifax, Nova Scotia

R. J. Weil, Assistant Professor of Psychiatry, Dalhousie University, Halifax, Nova Scotia

THE BUMPS

If on the street you chance to meet
A man who's pale and thin,
With sticking-plaster on his face
From forehead to chin,
And his whole frame shakes at the steps he takes,
And his head is full of lumps;
You may surmise, he's one of the guys
Who works down where she bumps.

When a man works in a place that "kicks"
He is up against no joke;
For he doesn't know when she might let go
(And she hits a nasty poke).
He holds his breath, he is scared to death;
Every noise he hears, he jumps;
For a man's in dread, fear he'll lose his head
When he's in a place that bumps.

At night when he crawls into his bed,
He can scarcely sleep a wink;
He'll lie and fret, for he's all upset,
And his nerves are on the blink.
If he falls asleep, out of bed he'll leap,
For he dreams he's at the "face";
It's a darned hard row, that a man must hoe,
When he's in a bumpy place.

L'Envoi

So when the old West Slope is bumping,
And all the boxes are "jumping";
When there's great big pebbles falling all around
Can you blame the guy that lingers,
With a smoke between his fingers,
In a kitchen where the roof is good and sound?

- A Minetown poem attributed to
Dannie Boutilier, "The Miners' Poet'

PREFACE

With the publication of this report, the thirteenth in the Disaster Study Series, another agent has been added to the types of events which have been previously analyzed in the series: a mine disaster. The pre-rescue period of entrapment for the victims was especially prolonged and stressful. This situation provided a rich and challenging opportunity for the staff that did the study. Their skills included a satisfactory balance of psychiatry, psychology, and sociology.

It is interesting to note that there is at least one distinct similarity--geographical--between this report and one of the pioneering social science studies of disasters, Samuel Prince's <u>Catastrophe and Social Change.</u> Both the 1958 mine "bump" as well as the 1917 ship explosion, which formed the basis of Prince's study, occurred in Nova Scotia and were analyzed by residents of that province. Of far greater significance is the contrast between 1917 and 1958 research concepts and techniques. The present study also attests to the ability of the 1958 research staff to construct new techniques in order to obtain a more systematic assessment of disaster behavior.

When the Academy-Research Council's work in the area of disaster research was initiated nearly a decade ago, more than 10,000 different accounts of disasters were extant. The vast majority of these had been provided by popular writers and journalists. The need to supplement this kind of information with systematic scientific research was central to the Academy-Research Council's decision to assist in the study of human behavior in disasters.

The present report is focused on a disaster which had many dramatic consequences and which has been the subject of a number of journalistic accounts. That there are unique differences between the way a journalist and a behavioral scientist examine and report the same phenomena is fairly obvious to most readers. Without disparaging the approach of the journalist, the present report provides an opportunity for illustrating some of the differences between the contributions which flow from these two sources.

Support for this study was made possible by the use of DRG funds from its National Institute of Mental Health and its Ford Foundation grants. Publication of this report does not necessarily imply agreement with every statement contained herein, either by the Disaster Research Group or by its sponsoring agencies.

George W. Baker
Technical Director
Disaster Research Group

ACKNOWLEDGMENTS

We would like to take this opportunity to express our thanks to the many persons and agencies who have contributed to this publication. The National Academy of Sciences--National Research Council made the project possible by its financial contribution. Harry B. Williams of its Disaster Research Group initiated the study and, along with Charles E. Fritz, gave the project staff the benefit of his specialized research experience in disaster. George W. Baker expertly guided the project from the stage of preliminary report to publication.

The collection of the data was made possible by the cooperation and assistance given by the people and officials of Minetown at a time when they might well have been preoccupied with their own difficulties. The miners and families who were interviewed and tested are to be specially commended for the willingness and graciousness with which they supplied information. The doctors associated with the Minetown Medical Clinic assisted in both personal and professional capacities. R. A. Burden of Minetown contributed invaluable assistance as special medical and technical consultant.

In the analysis of the data, the project staff is indebted for the contributions of Hugh Vincent, Psychologist, D. V. A. Hospital, David Doig, Psychologist, Victoria General Hospital; Solomon Hirsch, Associate Professor of Psychiatry, Dalhousie University, all of Halifax.

Throughout the study, special assistance was rendered by the Nova Scotia Legislative Library, and the Minetown and Nova Scotia Mental Health Associations. Dalhousie University, Halifax, provided office space and equipment.

J. W. Clark and William H. James of the Psychology Department, Dalhousie University, acted as editorial assistants in the preparation of this report. Their dedicated interest and efforts contributed greatly to the monograph. H. J. O'Gorman, Jr., Hunter College, New York, made many valuable suggestions. Dwight W. Chapman, Luisa F. Marshall, and Jeannette F. Rayner, members of the Disaster Research Group staff, and John H. Rohrer, consultant to the Group provided important critical comment. Mark J Nearman, also of the Disaster Research Group staff, gave splendid assistance in editorial work.

H. D. B.
R. A. L.

CONTENTS

List of Tables. xv

List of Figures xvi

Chapter 1   Methodology. 1

Chapter 2   Background of the Disaster. 7

Chapter 3   Surface Reactions and Rescue 15

Chapter 4   Behavior of the Trapped Miners:
             A Descriptive Account 35

Chapter 5   Behavior of the Trapped Miners:
             A Quantitative Analysis. 67

Chapter 6   Psychological Data on Trapped and
             Nontrapped Miners 83

Chapter 7   The Relationship Between Psychological
             Data and Initiations 99

Chapter 8   Psychological and Behavioral Analysis
             of Selected Individuals 109

Chapter 9   Evaluation and Summary 131

Appendix A: The Disaster Services in Minetown 141

Appendix B: Medical Findings on the Condition of
             the Trapped Miners 149

Appendix C: The Sentence Completion Test 153

References Cited 155

LIST OF TABLES

1. Three Groups of Wives Having Different Levels of Involvement: Post-impact Response by Per Cent of Group 26

2. Three Groups of Wives Having Different Levels of Involvement: Waiting Period Response by Per Cent of Group 28

3. Psychophysiological Symptoms During Entrapment: Group of Twelve 61

4. Psychophysiological Symptoms During Entrapment: Group of Six 62

5. Rank Order by Initiation Perception, "I" Initiation, and Group Evaluation Scores: Group of Six 70

6. Rank Order by Initiation Perception, "I" Initiation, and Group Evaluation Scores: Group of Twelve 71

7. Rank Order by Group Evaluation Score in the Escape Period, the Pinned Miner Episode, and the Survival Period: Group of Six 74

8. Rank Order by "I" Initiation Score in the Escape Period, the Pinned Miner Episode, and the Survival Period: Group of Six 75

9. Rank Order by Initiation Perception Score in the Escape Period, the Pinned Miner Episode, and the Survival Period: Group of Six 75

10. Rank Order by Group Evaluation Score in the Escape Period and the Survival Period: Group of Twelve 76

11. Rank Order by "I" Initiation Score in the Escape Period and the Survival Period: Group of Twelve ... 77

12. Rank Order by Initiation Perception Score in the Escape Period and the Survival Period: Group of Twelve 77

13. Comparison of Mean Intelligence and Personality Test Scores for Trapped and Nontrapped miners. 88

14. Mean Psychological Test Scores of Younger and Older Trapped Miners, and Younger and Older Nontrapped Miners 89

15. Correlation Matrix for Trapped Miners (N = 19) 92

16. Correlation Matrix for Nontrapped Miners (N = 12). 93

17. Mean Intelligence and Personality Test Scores of Eight-Day Group and Six-Day Group 95

18. Rank Order Correlations Between Test Scores and the Three Measures of Initiation for the Escape and Survival Periods Separately and Combined (N = 18) 100

19. Surface Disaster Relief Services 142

20. Distribution by Category of Physical and Emotional Condition and Treatment of Trapped Miners 150

21. The Physical and Emotional Condition and Treatment of the Trapped Miners by Code Category . . 152

LIST OF FIGURES

1. Diagram of No. 2 Mine Showing Area of Bump: Vertical Section 9

2. A Typical Mining Area Cross Section 10

3. Location of the Group of Twelve While Trapped: Ground Plan 36

4. Location of the Group of Six, the Pinned Man, and the Semi-isolated Man While Trapped: Ground Plan 38

CHAPTER 1

METHODOLOGY

This monograph examines individual and group behavior in a coal mine disaster that killed 75 miners and trapped 19 more underground from 6 1/2 to 8 1/2 days. The emphasis is on the men who were trapped--their behavior during entrapment, their physical and particularly their psychological state after rescue. More briefly considered are the rescue work, the miners who were not working at the time of the disaster, the wives of the trapped miners, and the general community setting of Minetown,* where the disaster occurred. Families of disaster fatalities were not interviewed.

The fundamental purpose of this study was to carry out basic research in human behavior. Specific social engineering aims were not of major concern and the investigators did not set themselves to answer a definite set of practical questions. Rather this project attempted to follow the lines of scientific method as closely as possible under emergency field conditions. As far as possible, standard procedures and techniques were used. Where no conventional technique was available for a particular condition, an exploratory one was attempted. A critical appraisal of these techniques might well assist future disaster research.

The Minetown disaster situation presented a number of advantages for research, although the investigators had little prior knowledge of the situation. The trapped miners obviously presented an important focus for the study--they had been in a real-life situation directly relevant to a large body of small group theory, laboratory studies, and natural group research, and they had experienced an event having implications for important problems in military and civilian life. Also, the situation limited the number of operating variables and provided groups that might be compared, facilitating a semi-experimental approach. Further the situation permitted the

---

*This pseudonym was adopted to protect the anonymity of respondents.

study, in an abnormal but natural setting, of several problems (leadership and morale, and their personality attributes) that have long intrigued social scientists

The entry of the research team into Minetown was accomplished through the auspices of a voluntary psychiatric community-relief service which had been instituted immediately following the disaster. Although this enabled the team to fulfill simultaneously two functions--research and therapy--it made relationships more complex, interviewer-respondent relationships were at times mixed with doctor-patient relationships. The clinical functions that, willy-nilly, team members were called on to perform perhaps slightly vitiated the scientific detachment and objectivity of the study as a whole (Eaton & Weil, 1951).

The data were gathered within the general theoretical framework of psychiatry, psychology, and sociology. Most of the data were quantified and, where feasible, analyzed by standard statistical procedures. Although the samples were not large, quantification was stressed in an attempt to exert rigid control on impressionistic observations.

Subjects and Procedures

The data for this study were collected from six groups of subjects.

(a) Nineteen trapped miners. These men were subdivided into three subgroups by the nature of their entrapment and its duration. 12 men who were trapped in one location for 6 1/2 days; 6 men who were located in a different place and were rescued after 8 1/2 days; and one man who was in semi-isolation for 8 1/2 days some distance from the subgroup of six.

(b) Twelve nontrapped miners. These men were selected from the miners who had worked on the day-shift during the day of the disaster. They were matched with the 19 trapped miners for age, education, religion, marital status, and type of work performed in the mine. Eight of the nontrapped miners had participated in rescue operations.

(c) Seventeen wives of the trapped miners.

(d) Eleven wives of the nontrapped miners.

(e) Ten personal service and professional personnel representing the services generally rendered in a small mining community, like Minetown. They were a nurse, a school superintendent, a school principal, a minister, a union official, a store proprietor, a bank manager, a medical doctor, a government official, and a mine manager.

(f) Seven wives of personal service and professional personnel.

All the data from these 76 subjects were collected by means of interviews and psychological tests carried on from 5 to 23 days after the miners had been rescued. The division of labor among the three members of the field research team was determined not only by the specialized interests of each member but also by the need to collect the data quickly and by the psychiatrist's previous service role in the community during rescue operations. The psychiatrist interviewed all miners, trapped and nontrapped. The sociologist interviewed the professional and personal service personnel of the community and all wives. The psychologist administered the psychological tests to the 19 miners who had been trapped and to the 12 nontrapped miners.

In the majority of cases, the psychiatrist interviewed a miner only once. This restriction was imposed by a benefactor's taking the trapped miners on a vacation from Minetown soon after their rescue. The psychiatrist's interviews were focused first on the individual's experience and behavior during the emergency period, that is, from the time of impact to the completion of rescue operations. For the trapped group, this involved the period of entrapment, for the nontrapped group, it covered the period of rescue operations. The second part of each interview covered the man's feelings and behavior since completion of rescue operations, and his view of the future. The final part of the interview obtained the man's life history. An attempt was made to use the nondirective method of interviewing, especially when obtaining the miner's account of experience in the disaster. Subjects were interviewed in the hospital or in their homes. All but five of the 31 interviews were recorded on tape and later transcribed. The psychiatrist made notes while conducting the five untaped interviews and dictated his report of the interview later the same day.

The sociologist interviewed the 45 wives and the personal service and professional personnel in their homes or at their place of work, using a prepared questionnaire form. The questions were focused on indications of preparedness for disaster, first reactions to disaster, and activity during the emergency and waiting periods. The interviews were conducted in a somewhat directive manner, with the respondents being allowed to elaborate when they wished. Their verbatim responses were written on the questionnaire.

The purpose of psychological tests in this investigation was to measure some of those functions that might be affected by entrapment underground, such as immediate memory, attention and concentration, visual-motor coordination, analytic thinking, handling of unstructured material, and attitude. The objectives, together with consideration of time limitations and the makeshift research conditions, decided the selection of the following tests.

(a) The Vocabulary, Digit Span, and Block Design subtests of the Wechsler Adult Intelligence Scale (Wechsler, 1955).

(b) Counting backwards from 20 to 1, and counting backwards from 100 by 7's.

(c) Bender-Gestalt drawing test (Bender, 1938).

(d) Rorschach ink blot test (Rorschach, 1942).

(e) Sentence Completion test (Appendix C).

The Sentence Completion test was constructed for the study expressly to elicit information about the subject's fears and anxieties, expressed needs; evaluation and expectation of self and others; and attitudes toward mining, bosses, leaders, religion, and the future.

The exigencies of the research situation, particularly the limitations of time and the miners' susceptibility to fatigue, made it necessary to implement modifications of the test battery in the field. The Vocabulary subtest was given to only 17 trapped miners and 3 nontrapped miners. The memory part of the Bender-Gestalt test was omitted after the first few subjects expressed considerable anxiety and took an undue amount of time trying to recall the figures. On the Rorschach, the inquiry, that is, the follow-up questions after the initial response, disturbed the first subjects, took a long time, and gave virtually no more information than was elicited by one or two general questions. For these reasons it was limited thereafter to two questions (a) "How much of the blot looks like...?" and (b)

"What about the blot makes it look like...?" The final modification was to give only five of the Rorschach cards, namely, I, III, VI, VIII, and X. This change may raise objections from clinicians, but the observance of the standard test method was precluded here because many of the miners were too tired to tackle all the Rorschach cards. Methodologically, this decision is justifiable in terms of reliability studies of ink blot cards (Dörken, 1950; Frank, 1939; Hertz, 1934). Moreover, in this study a control group was used for comparative purposes.

Treatment of Data

The interview responses of the wives and nontrapped miners that were relevant to particular questions were coded and counted. The section dealing with their experiences and behavior also makes use of descriptive quotes and anecdotes (Chapter 3). Those parts of the trapped miners' interviews that were about their experience and behavior underground and that were essentially nondirective were subjected to a content analysis in terms of "initiations." This made it possible to discuss the group interaction of the trapped miners on a comparative basis (Chapter 5). The psychological test data were all quantified. As an additional means of quantifying each miner's personality assets, two psychiatrists analyzed each interview for several personality attributes. These attributes were rated and the sum of the ratings was taken as the man's psychiatric ego strength score. Psychiatric ego strength and psychological test data were then analyzed within and between groups by means of statistical techniques (Chapter 6). In appraising the individual trapped miners in terms of predisaster history, psychological makeup, and role behaviors while trapped, quantitative information constituted the primary basis for inferences.

CHAPTER 2

BACKGROUND OF THE DISASTER

At 8·05 p.m. on October 23, 1958, an underground upheaval devastated No. 2 mine in Minetown. Of the 174 men who were in the mine at the time, 74 were fatally injured, 100 were rescued, and one of these died of injuries two weeks later. The day after the last body was removed from the mine, the company announced that it would not reopen the mine, and so brought ninety years of mining operations to an end in this coal mining town. This study is focused on the subjects' behavior during the emergency period.*

The Community

At the time of the disaster Minetown had a population of 7,138 (Dominion Bureau of Statistics, 1953). The majority were Protestants of Anglo-Saxon background, with a fair-sized Roman Catholic minority. There were also a few Negroes. Minetown has always been a town of families, with no tradition of transient immigrant labor or dormitory coal camps such as Luntz (1958) observed in Coaltown. Eighty per cent of the town's families were home owners. Although Minetowners thought of themselves as belonging to a stable community with a high degree of occupational inheritance, 35 per cent of the population had moved into Minetown within the current generation. For the most part they came from the immediate agricultural hinterland. About 19 per cent had been in Minetown for three generations.

The town is connected to the main railway line by a four-mile spur line. It is also at the junction of two highways. The surrounding mixed agricultural land and rocky, wooded terrain provide excellent hunting and fishing. The nearest larger town is 55 miles away.

Minetown has been a one-industry town. the economy has been almost exclusively dependent on coal mining. At one time there had

---

*The long-term implications of the disaster for individuals, groups, and the community as a whole form a separate study. It has been conducted by largely the same staff that conducted this study in 1958.

7

been several mines, but at the time of the 1958 disaster only No. 2 mine was still working, all the others having closed for reasons of safety and economy. Number 2 mine employed one thousand men. Only twenty people were employed in other primary production industries. All other employed community members were engaged in providing personal goods and services to mining personnel and their families.

The Mine

Number 2 mine is one of the deepest coal mines in the world. Figure 1 is a diagram of it. The mine shaft entered the earth at a slope of approximately 30 degrees. At a distance of 11,400 feet (called the 11,400 foot level), and a vertical depth of 4,348 feet, the slope levelled out to about 12 degrees. The main slope extended to the 7,800 foot level and was joined by tunnels to the back slope.

The coal was bituminous, lying in seams 8 1/2 to 9 feet high. The coal was raised and the men carried in and out of the mine by means of an outside electric hoist that lowered one "rake," or group of boxes and trollies, while another was being hoisted. At the top of the back slope a second electric hoist lowered and hoisted rakes from the four working levels on the east side of the mine. the 12,600, 13,000, 13,400, and 13,800 foot levels. A short level at 14,200 feet was maintained for circulating the air for ventilating the mine.

Air pipes conducted compressed air to the machines that were used on the coal faces, or walls. The circuit for ventilation was down the back slope to the 14,200 foot level and out the 12,600 foot level. The temperature in the mine rarely fell below 80° Fahrenheit and rose only slightly higher in some locations.

The coal was mined toward the slope by the retreating longwall method. This means that as the coal was removed from the walls a new row of "packs," i.e., roof supports, was built behind the worker in the recently evacuated area and the last row removed so that the roof was allowed to fall in behind the new working area. This unsupported roof area was called in Minetown the "waste," and in some mining communities the "gob." As the miner cut coal out of the wall, behind him, in order, were the pan line, an area with a roof supported by packs, and the waste without roof support, sometimes fallen in, sometimes still holding (see Figure 2). The men moved up and down the travelling road between the rows of packs parallel to the wall.

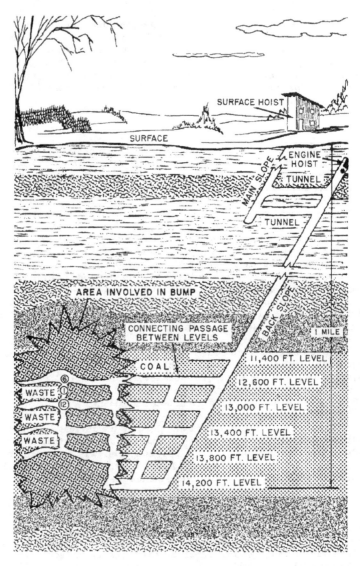

Figure 1. Diagram of No. 2 Mine Showing Area of Bump: Vertical Section

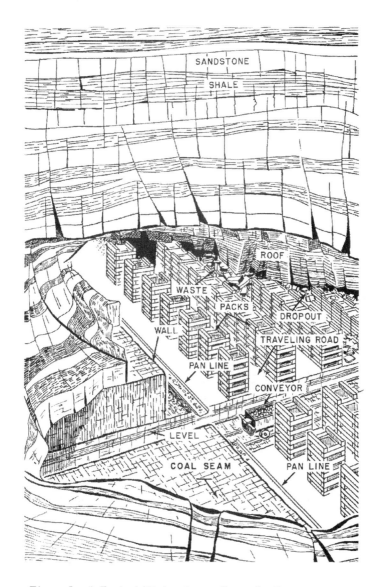

Figure 2. A Typical Mining Area: Cross Section

When taking coal out of the wall, the miner was higher than the level (see Figure 2) so that the coal ran down onto a pan line, or metal trough. This trough was shaken longitudinally by the pan engine to slide the coal down onto conveyors which moved it out to the loader and into coal boxes at the end of the level. These coal boxes were then pulled along the levels on tracks by stationary engines and cables and finally up the slopes to the surface by the hoisting engines.

The men worked on three nonrotating shifts. Two hundred worked on the surface. Eight hundred worked underground, each man carrying his lunch pail and a large can of fresh water of greater quantity than would usually be required for one shift. There was a wide division of labor. Of the men underground, 270 were "contract" miners, those who actually dug the coal from the walls. Before becoming miners, they had to be physically fit and have experience in the mine, special training, and certification. They dug coal in groups of about 30 on a group contract basis and were paid according to the amount of coal reaching the surface from their walls. These contract miners were usually in the mine for a ten-hour shift. This permitted them to work an eight-hour day at the coal face. They were not paid for travelling the long distance from mine portal to coal wall. Total earnings were divided among the members of the group. Contract miners earned higher pay than the other workers, particularly if they belonged to a high-producing, cooperative group.

The remaining 530 underground men were "datal" laborers. They worked for an eight-hour day, portal to portal, for a fixed day's pay. The datal laborers supported the contract miners with many different services, including building packs to support the roof, operating hoists and machines, and cleaning up rubble. A small group, called "shift men," cleared the mine of any rubble and maintained and checked the machinery on the night shift when coal was not mined. The datal laborers, usually working in pairs, were much more mobile in the mine than the contract workers. Their skills, like all skills in the mine, required few formal educational qualifications. They were learned on the job and were not readily transferable to jobs outside a coal mine.

Mining Dangers

Safety precautions were necessary because coal mining is a dangerous occupation. Minetown miners had to work with danger of three types. First, there were humanly preventable mine explosions, which brought death on a mass scale two major mine explosions, one in 1891 and one in 1956, had killed 164 Minetown miners

The second type of danger the miners faced continuously was the threat of individual injuries or death. Individual accidents in the mines had killed 176 men, an average of two deaths per year. Of these, 96 were attributable to human error; 80 were due to uncontrollable and unpredictable forces of nature.* There had been only 13 years free of fatal accidents in the history of Minetown. Seven of these were prior to 1900. These figures do not take account of deaths caused by occupational disease due to methane, carbon monoxide, dust particles, etc., or by the many serious injuries.

Individual accidents and explosions are present in all mines, but a third threat, the "bump," is found only in certain mining areas. A bump may be described as a sudden bursting of the coal or of the strata immediately in contact with it. It is usually accompanied by a loud report and by ground tremors, which are sometimes felt at great distances from the point of origin. The primary cause of a bump in a coal mine is believed to be due to pressure that is exerted on the coal seam and its underlying strata by the weight of the overlying mass. This may result from mining operations, particularly when mining is at great depths. (Report..., 1959). Bumps are uncontrollable and completely unpredictable, and in Minetown they occurred frequently. In No. 2 mine, they had been a major concern to workers, the company, the government, and geologists alike.

Bumps have buried or killed whole groups of men and dislodged stones that have required the combined efforts of five or six men to move. Other bumps have shattered the coal wall, and the coal has "run" in tons, injuring or burying workers as they dug to reach buried comrades. Bigger bumps have resulted in major shifts of the strata with underground roadways being heaved against the roof. Trollies of several tons weight have been crumpled and driven into the rock and coal.

Preventive measures suggested by world experts, as well as by local folklore, have had little effect. The first recorded bump in Minetown occurred in 1917 at the 4,700 foot level. Subsequently, small bumps have occurred frequently, and periodically there have been large or "district" bumps. Attempts to counteract the danger

---

*Figures especially compiled by the Legislative Librarian for this study from the annual reports of the Department of Mines from 1876-1956.

led first to a modification of "room and pillar extraction," then, in 1925, to the adoption of "retreating longwall mining." Later, the levels were widened, additional stonefilled packs built, and many other modifications made. Some deeper levels were abandoned because of severe bumping. None of the modifications have eliminated the occurrence of bumps. In all, some 500 bumps have been recorded in No. 2 mine. Between March and October, 1958, underground personnel recorded 17 bumps. Although these 17 were all small bumps, 50 men had been injured, one fatally (Report..., 1959).

Safety and Rescue Procedures

Each group of contract miners had an "official," who was a fellow contract worker designated by the company to look after the safety of the group. Checking for gas was one of his responsibilities. Regulations required the official to evacuate the area if the methane content reached 2.5 per cent.

A few contract miners with long experience and familiarity with the mine became "dreagermen." Dreagermen were specially trained in rescue work and were the first to enter the mine in the event of a mishap. They wore masks with a two-hour oxygen supply, and carried 38 pounds of special equipment that allowed them to work in gas-filled places. This job provided high prestige, but no extra pay. In all levels of the mine for each shift, the company had designated miners to be responsible for directing rescue work and cleaning up when an accident occurred.

All underground workers, at the start of their shift, changed into their pit clothes in the wash house. At the lamp cabin they received their lamps in exchange for a metal disc. The lamps fitted on their hats and were powered with wet-cell batteries. They gave light for about twelve hours without recharge and could be turned off and on by the miners to help preserve the life of the battery. These were the only lights in the mine, so the men wore them continuously. The lamps were numbered with the men's check numbers, as were their lamp discs, so that in an emergency a tally of personnel in the mine could be made rapidly by counting the discs on the board at the lamp cabin. A badly crushed body could be identified by the number stamped on the lamp battery strapped to his belt.

An "official" in the mine placed each man in a particular location and noted in his record book where the man was supposed to be. This was not always accurate as some jobs required a man to circulate considerably in the mine. The official's record book, however, did provide another check on the location of a miner.

The 1956 Explosion

The last major disaster in Minetown prior to the 1958 bump had taken place two years before, when an underground coal dust explosion and fire trapped 127 men in No. 4 mine. Two dreagermen lost their lives in the early part of the rescue. Within four days, 88 men were rescued and the remaining 39 men, who were known to be dead, were sealed inside the mine in order to control the fire and their bodies were recovered later. Number 4 was not used again, leaving No. 2 mine as the only working mine.

Immediately after the 1956 explosion occurred, the disaster organization of the mine was put into action. Dreagermen entered the mine and operated in smoke and carbon monoxide. Additional dreagermen came to Minetown from other mining centers. "Barefaced" miners, without masks or oxygen, carried out underground rescue work in unpolluted areas. On the surface, the hospital was prepared for emergency service. Relief organizations entered the town bringing supplies and special skills, set up and staffed canteens, auxiliary hospitals, first aid centers (one of which was staffed by a psychiatric team), and other services. Townspeople manned local organizations. Newsmen and commentators covered the story from the pit head. All available space was taken up in the community.

Although the 1958 bump created a different type of mine disaster, many of the same miners, dreagermen, and rescuers were involved. It affected the same town and directly reached into many of the same homes. Many of the same relief organizations with the same personnel rushed to the scene and took up the same locations and duties. The 1956 explosion provided many patterns that were followed in 1958. Although each disaster has unique aspects, and, consequently, not all behavior was affected by the earlier precedent, the fact that so many people had been involved so recently in large-scale emergency action affected many facets of the 1958 operations.

CHAPTER 3

SURFACE REACTIONS AND RESCUE

The bump of October 23, 1958, was the most severe in the history of Minetown. The shock was registered on a seismograph 800 miles away. The bump was felt as a distinct tremor in communities 15 to 20 miles away. In Minetown, almost everyone felt the ground tremor.

A miner who had been on the day shift was walking down Minetown's Linden Street at the moment of impact. He reported:

> I thought some kids put a bomb under Jim Brown's house. I said, "What in the hell's that?" and he said, "I don't know!" And then a neighbor came out and she hollered, "What was that?" and I said, "I don't know." So I ran home, and I said, "Mary, did you feel that bump?" And she said, "It knocked me off the couch." And I said, "Lord God! it's the whole three walls must have went!" and I ran to the pit.

Four thousand feet underground, one of the 174 miners working in the mine had just stepped over the seat of the trip car when the bump rocked the mine

> I can't recall how long I was out, it might have been a day or an hour or it could have been five minutes. I really don't know. When I came to, everything was right to the roof, the rails and the unit that I was operating was just a mass of steel. It was all smashed and drove right into the pack. Nearly everything was right to the roof. I don't know how I got out myself; there was a space there about six or eight inches in height. I must have got through it, but I don't remember coming through it myself.

In one of the homes in Minetown, the wife of a miner who was at work in the mine was sitting alone in her dining room watching television:

I knew exactly what happened. It is hard to explain, but it just felt like the whole world hit the house; it was terrific. I knew it was a bump and I thought of my husband in the mine. I said right out loud, "My God above, it happened!" My husband had always said, "If anything ever goes wrong at the pit, don't go near the mine." So I immediately went to my sister's next door. When we heard it over the radio I was worried about my daughter who is a stenographer in Yorktown. She would hear the radio and be very worried about her father. At that time, I expected word about the men by midnight, and I wanted my two daughters home then.

Several streets away, a nurse felt the impact and after checking everything in the house decided that it must be a bump. She was terrified at the thought of the seriousness of the impact. She cried. She went to her neighbors to confirm her conclusion. From there she went to the pit head, heard a report that the men were all right, and returned to her home. Later she went to bed, only to get up again, dress, recheck the seriousness of the bump, and subsequently report to the hospital.

Although the main focus of this study is on the miners trapped in the mine, brief consideration will be given to the rescuers, rescue operations, and the waiting miners' wives. First, the off-shift miners who were in the town on the evening of the bump will be discussed.

Off-shift Miners

It has been noted in many disaster studies that if persons or groups have no forewarning or expectation of the impact of a disaster, they initially tend to define or interpret the event in terms of cues that are familiar to them (Committee on Disaster Studies, 1955). This tendency to assimilate disaster cues to normal or usual expectations has been reported in virtually every disaster study (Fritz & Rayner, 1955; Marks & Fritz, 1954, Spiegel, 1957; Young, 1953). Minetown serves as an unusual situation in which to observe the reaction to first impact because the citizens and particularly the miners themselves, were skilled in mine lore and the signs and signals of tragedy. A bump of such great intensity, however, was a unique experience for identification and interpretation against these well-known cues.

Twelve miners who were off shift at the time of the bump were interviewed. None was alone when the bump occurred. Six of the 12 said they immediately recognized the tremor as a bump. Although they thus judged correctly, they talked to the interviewer in similes that they thought the interviewer would understand "It felt like a big truck hit the house," or, "It felt like the house had fallen off its foundation."

Five thought the impact was something other than a bump: a bomb, a fallen stool, children upstairs, or a truck hitting the house. The other miner from the 12 interviewed was driving his car at the time and was unaware that the impact had taken place.

Clearly, although all Minetowners were at least knowledgeable in mining matters, it was primarily those that had relatives in the mine at the time of impact who immediately interpreted the impact as a bump. Five of the 6 miners that immediately interpreted the impact as a bump had relatives in the mine at the time, 15 of the 17 wives that had husbands in the mine at the time did likewise. On the other hand, of the 6 off-shift miners that did not have relatives in the mine at the time, all interpreted the impact as something other than a bump, as did half of the 18 wives whose husbands were not in the mine at the time. This direct personal involvement rather than a general background and orientation seems largely to account for the difference. However, an important question remains unanswered: why did none of the non-personally involved miners, all of whom were experienced with bumps, not identify the event correctly?

The degree of general background and orientation may be a factor in the apparent accuracy of interpretation of a disaster event. Whereas 62 per cent of all subjects in this study made a specific and correct interpretation, only 10 per cent in Killian's (1956) sample made a comparable interpretation.

Virtually every disaster account makes reference to the jamming of telephone facilities when a slightly increased percentage of the population attempts to use the system simultaneously (Fritz & Mathewson, 1957). Minetown was no exception. A number of people reported that they were unable to complete telephone calls successfully. But while all the off-shift miners confirmed or redefined their original conclusions about the nature of the tremor within a few minutes, only one reported using the telephone. The majority carried out their social confirmation in the group in which they found themselves at the time of impact. In addition, about half of the sample of miners confirmed their first impressions with a larger group on the street or with neighbors.

Rumors are present in virtually every disaster situation (Prasad, 1935). The situation in the first few minutes after the impact was particularly susceptible to rumor, for rumor increases with the importance of the subject to the individuals concerned and with the ambiguity of evidence pertaining to the topic at issue (Allport & Postman, 1947). One of the off-shift miners had his particular interpretation of the bump verified in informal discussion.

> In the street they told me that the bump had been in the waste, and no one was hurt. If it was in the waste, that's happened so often it just doesn't mean anything.

This was accepted by this experienced miner long before any accurate information could possibly be available. Things people expected to happen were reported as having happened (cf., Larson, 1954).

All but one of the miners interviewed went to the mine within a few minutes of the impact. This particular case of convergence at a disaster site illustrates a number of separate categories of convergence as specified by Fritz and Mathewson (1957). The off-shift miners converged upon the mine, not only for information and evaluation of the bump, but also as helpers responding to the informal social code of miners that they assist in rescue operations as soon as a disaster occurs. They were helpers on an informal voluntary system and were simultaneously integrated into the existing formal structure of rescue operations. In addition, each was anxious about relatives or friends.

Miners who volunteered for rescue work were the supporters of families, and in volunteering they were risking their families' support as well as their lives. Yet nowhere in the inverviews was there any suggestion of the conflicting loyalties reported in other disaster rescue studies (Killian, 1952). Rescue work was regarded as a duty to friends, co-workers, and the occupation; it was part of the discipline of their work. Some had relatives in the mine. Others had themselves been saved in previous mine disasters. The majority had a general identification with their fellow miners by virtue of their common occupation and the knowledge that they themselves might face a similar situation in the future (cf., Marks & Fritz, 1954).

If the loyalties to these various groups and ideals had been weighed against loyalty to the family, it would be expected that the multiple loyalties symbolized by entering the mine would have far outweighed the responsibility to the family. There was no evidence, however, that the issues were weighed, apparently because the family itself had loyalty to the miners' norms. The wives had relatives, friends, and husbands of friends trapped in the mine, and the wives had to live in a community in which behavior was judged in part by the males' conformity to the code. In no interview could any reservation concerning a husband's decision on the part of the miners' wives be found. Rather than conflicting group norms, Minetown rescue behavior illustrates the reinforcement of common values shared by multiple groups.

In the sample of 12 off-shift miners, 8 took part in rescue operations, and of these 8, 5 suggested that it was the "natural" thing to do--implying the informal code: "My idea was to get down and see what I could do below." One miner also spoke specifically of relatives "I worked till it was done, till it was over--I brought up my brother." On the other hand, three were officially approached and specifically assigned "We was asked to be there--we've always been on what they call the wrecking crew." But the informal social code of rescue, social pressures, and knowledge of appropriate role did not mean that a rescuer was able to fulfill it. One of the men had a physical disability that precluded rescue work, one was unable to participate because of "nervous" reasons; and one gave no explanation. The fourth stayed within the code; he volunteered but was not accepted:

> I heard they wanted volunteers to go down, so about fifteen minutes after, I went over. And he said, "Well, the amount of men we wanted we just got." So I said, "There's no need of my services" and went back and never bothered them no more.

The miner who could not face the task because of nervousness discussed the breaking of his own expectations and the group code in this way:

> I tried to go down, but I couldn't--I was in rescue work before, and I don't know if I lost my nerve or what, but I just couldn't. I couldn't make it, and that's all. What I thought about was if I went down there and went all to pieces I would have to have two men bring me up, and where those two men could continue on and do something valuable. That's the way I figured it.

One of the men who was assigned reported reluctance when he was telephoned.

When they did call on me, it came as quite a shock because I figured they had lots of men and they wouldn't need me. And when they did, it left kind of a hollow feeling in my stomach I wouldn't want to have that experience again, not even going down as a rescuer. I don't think I'd volunteer again really.

The miners, however, were not all forced to make a public decision if they did not wish to do rescue work. Few were formally recruited or telephoned personally by those responsible for mine rescue crews Many more than the 60 or 70 rescuers needed on each shift were available, therefore, it is impossible to determine just how many did not report or volunteer on their respective shifts as a personal choice.

Rescue Operations

The Royal Commission (Report..., 1959), stated that it is a matter of commendation that within minutes of the knowledge of the bump rescue operations were set in motion. Immediately after the mine manager felt the bump at 8.05 p.m., he telephoned the chain runner stationed on the 7,800 foot level. He was informed that everything was in order on the main slope, however, no communication was available with the sections below the 7,800 foot level. Arrangements were made to hoist the coal cars on the slopes to clear the way so that "man rakes" could be lowered with rescue personnel. Fans were checked and found in good running order, but power was not available in large sections of the mine.

Mine exploration and rescue of this type is a specialized and highly intricate operation. Unlike other similar situations in which outside direction was needed (Gordon & Raymond, 1952), the entire operation was meticulously carried out by the Minetown company and its employees. No direction from outside the community was required.

The surface plant was placed on the disaster organization plan. The rescue station was opened, and the gas-analyzing apparatus prepared to run complete determinations. Senior officials were notified and, after consultation with the rescue superintendent, arrangements were made for additional dreagermen and equipment from other coal communities.

The dreagermen arrived at the mine, and many miners who had come to the pit head volunteered for rescue work as "barefaced miners." Barefaced miners work in gas-free areas on an eight-hour shift and, unencumbered by masks and oxygen equipment, they can work harder and in more difficult positions than the dreagermen.

At 8:40 p.m., the mine manager entered the mine with 13 men including representatives of the Union and the Department of Mines. First, communication was restored to the 13,400 foot level. It was reported that the inside end of the 13,000, 13,400, and 13,800 foot levels was closed off outside the walls and no word had been received from any of the men on the wall faces

About 9 30 p.m., the rescue party reached the 13,400 foot level where it found the level blocked by methane gas. The party then made its way to the 13,800 foot level. One badly injured man was found. Thirty additional men, including a doctor, were called down to aid other possible injured. At 11 30 p.m., 12 live men and a fatally injured man were found. The live men were sent to the surface immediately. At about 1:30 a.m., the bottom of the 13,400 foot level wall was reached and 11 men were rescued and sent to the surface.

Of the 174 men in the mine, 81 were rescued or escaped within 24 hours. Nineteen of these were injured, some severely. This bump was unique in that it occurred at the coal face, or wall, and not in the levels where bumps usually struck. As a consequence, most of the 81 men who came out alive had been in the levels and slopes at the time of the shock. However, some of them had been in areas on the wall where destruction was not complete. It was not known whether anyone had survived at the top of the 13,400 foot level wall and along the 13,000 foot level wall, since all entrances to these areas were blocked by solid falls of stone, coal, and debris.

Rescue work proceeded without pause, directed by the mine manager. After the first 24 hours, between 150 and 200 miners, divided into their regular shifts, worked at rescue. They operated in teams, each team with a complete range of skills. The four local dreagermen teams were reinforced by 48 dreagermen from other coal mining towns. Three doctors entered the mine from time to time when necessary.

This underground rescue operation was supported at the pit head by a multiplicity of special services that had converged upon Minetown from a radius of 150 miles. At the time of the bump, it was not known how many injured men would have to be cared for. Emergency and relief agencies dispatched teams and equipment to the town, army and civil defense units moved in, first aid stations were set up at the pit head; mobile kitchens were opened, supplies of oxygen were rushed in; a community psychiatric service was staffed around the clock. The majority of these services and personnel had rushed to Minetown for the 1956 explosion, and this previous experience patterned their spatial location and the division of labor during the 1958 bump with a minimum of confusion. The list of the main services and an appraisal of their effectiveness is found in Appendix A.

The Royal Commission (Report..., 1959) reports that rescue operations were carried out under almost impossible conditions and the rescue of a large number of men was brought about by the continued and persistent efforts on the part of those engaged in the rescue work. Under normal operations there was a height of over eight feet along the wall faces and levels, but the violent upthrust of the floor reduced the height to about three feet and, at intervals, to much less. Coal ribs had burst and pushed into the level approaches, completely closing them. In addition to this, the ventilation was disrupted. Roads and equipment were thrust against the roof creating additional obstacles that retarded and handicapped the rescue operations.

Dreagermen continued to work at the restoration of ventilation for the 12,600, 13,000 and 13,400 foot levels. The barefaced miners had to exercise great care in their work, as it was feared that the removal of any coal under stress could result in another bump. The men engaged in rescue work were in constant danger. They were aware of this danger, but did not acknowledge it to one another:

> We done all the hollering and talking and chewing--everything we could think of to one another--to keep our minds off the bumps. We knew what we was under, we knew what we had to go through. See? We knew that.

To accomplish entry in levels and along wall faces, debris had to be removed by makeshift methods. Only one man at a time could dig at the "face" of the tunnel, and even then, the handle of the pick had to be cut off. As he dug, another man scraped the coal back with his hands, and a third, using a shovel without a handle, put coal in water buckets. The buckets were passed back by a bucket brigade from man to man in the narrow tunnel:

> If it hadda been one man there alone--working there--he couldn'ta done nothing. But four of us spelled one another off. And like that, I'd go in there first on my stomach and dig and plow--maybe for two minutes, maybe for five, maybe ten--no longer than ten at any time. There wasn't the right air for you--anyway to begin with. And then it was so dusty. The minute you stuck a pick in it you got

that dust, and sometimes you could hardly see. I'd
come out and the other fellow'd crawl in over my back
and go in and dig. And I'd sit there on the high side
and rest. But the four of us kept going around in a
circle like that.

Despite the danger and discomfort, most rescuers reported
that their greatest difficulties arose from handling crushed and decomposing bodies.

You had to dig each body out--maybe strike a boot
toe first, maybe strike a cap first. You can imagine--
down here in a hole--that high--sometimes lower than
that, digging, pulling them out. My buddy took so sick
he had to go home. I think maybe my stomach is just a
little stronger than his--could take a little more. The
smell was awful, we had that spray stuff which did help
a lot but didn't kill it all We had rubber gloves on,
but even so we had to handle every body. I used to just
take sick and throw up and go back at it again--there
were four or five of us in that crew. Well, the rest of
them were real good men. Some fellows just to look
at them would turn sick, and some fellows couldn't eat.

There was great admiration expressed for miners who did not
react physiologically or psychologically to the work and the odor.

Many rescuers made a distinction among an unidentified body,
the body of a friend, and the body of a relative. One respondent
noted that if he did not see the body of a relative, the next of kin
could not ask him about it. Another expressed the difference between friends and relatives in this way:

I was just as glad that somebody else got my cousin
(the dead body) because those were all my friends, everybody down there was friends. Well, it's bad enough to
run into them when you're digging, but when you run into <u>one of your own</u>, it hurts worse, don't it?

Nevertheless, the men worked regular eight-hour shifts, and
within each shift the pick never stopped, a new man taking the pick
from the exhausted man's hand. Minor coal bumps occurred as the
rescuers dug their way forward. Gas forming from the coal was removed and ventilation to the tunnel was provided by a suction fan in
the end of a pipe the length of the tunnel.

From Friday, October 24, until Wednesday, October 29, the rescuers worked without finding any men alive. Bodies were recovered one or two at a time and brought to the surface with the change of shifts. On Monday, the manager issued a press release in which he stated that there was very little hope that more men would be found alive. He did make one reservation there was a possibility of life at the 13,000 foot level wall, which had not yet been reached by the rescuers.

Work was started on this wall, and it was found that the inside of the 13,000 foot level was completely closed. Opening it involved burrowing with the bucket brigade a distance of 401 feet. The wreckage and rubble were so great that it took over 48 hours of continuous work in relays to advance 328 feet. At 2:30 p.m., Wednesday, October 29, they reached the six-inch air line in the level and found that it was parted. While an air sample was being taken in the pipe, the voices of 12 men trapped at the bottom of the 13,000 foot level wall were heard. The remaining distance of 73 feet was gouged out in 12 hours. During this period, a copper tube one-half inch in diameter was pushed through the air line. Under a doctor's supervision, liquids were pumped through the tube to the trapped men.

When the men were reached at 2.30 a.m., Thursday, October 30, they were found to be in reasonably good condition. One of the men had a broken leg and another was suffering from an internal injury. The last of the 12 men rescued at this location reached the surface at 5.12 a.m., Thursday, October 30, after 6 1/2 days of entrapment.

Immediately after the 12 men had been taken to the surface, the 13,000 foot level wall was explored uphill for a distance of 160 feet. At this point, the wall was blocked by rock and coal and an opening had to be made through the midwalls and the waste toward the top of the wall. This work continued until Saturday, November 1, when, at 4.42 a.m., the rescuers found a lone man. This man was trapped in a small hole and was in such a weakened condition that he could give no information. In the words of a rescuer,

> He was in a little wee cubbyhole not higher than the arm of a chair. He was down, laying by the pans, and all this waste hanging up behind him. If it had fell, it would have buried him up.

A small opening out of the isolated man's cubicle was enlarged, and after crawling forward a few hundred feet the rescuers found six more men alive. The six men were first given water and then hot soup and coffee after suitable intervals. The trapped men crawled down the wall, where they were placed on stretchers. These men arrived at the surface at 9:15 a.m., Saturday, November 1, after 8 1/2 days in the mine.

The exploration continued until 8:25 p.m., Thursday, November 6, exactly two weeks after the bump. By that time the last of the 74 dead bodies had been discovered and taken to the surface.

Behavior of Wives in the Community

The 17 wives of the trapped miners were interviewed about their behavior during the period of impact and the long, subsequent waiting period. To check if this behavior was typical of all wives in the community rather than specific to wives of trapped miners, two small samples of less personally involved wives were interviewed: a group of 11 wives whose husbands were off-shift miners (many were rescuers), and a group of 7 wives of professional and personal service personnel who were least involved.

The Night of the Bump

The distribution and frequencies of a number of early reactions to the bump by the three samples of wives are shown in Table 1. All but 2 of the 17 wives of trapped miners said they knew immediately that the "shaking of the house" was a bump. The two exceptions were a woman who was driving a car and another who thought that "it was just the TV aerial falling down again."

Even though most of them knew immediately that there had been a serious mine disaster when they felt the tremor, all wives reported some kind of rapid social confirmation of the event. A half confirmed the bump through neighbors and a half through relatives

Other studies have found that individuals under stress tend to form groups spontaneously. It has been suggested that catastrophe produces strong feelings of dependency (Tyhurst, 1954a), and that groups seem to have the function of explaining and redefining an ambiguous but threatening situation (Larson, 1954). While all of the wives of trapped miners joined groups, fewer than 30 per cent of the wives less directly involved did so. Certainly many of the less involved wives had their husbands nearby to supply both information and support.

TABLE 1

Three Groups of Wives Having Different Levels of Involvement:
Post-impact Response by Per Cent of Group

Response	Wives of trapped miners (N = 17)	Wives of off-shift miners (N = 11)	Wives of professional and service personnel (N = 7)	All wives (N = 35)
Interpreted impact as bump	88	55	43	69
Had someone in family go to pit head	41	73	85	60
Received confirmation from neighbors, friends, or relatives	100	27	29	63
Did not go to bed on first night	94	73	13	71
Tried to use telephone	41	55	13	40
Left home to visit someone during first night	59	64	71	63
Had visitors on first night	65	73	57	66

The attainment of both information and social management of the situation was accomplished by face-to-face contacts, rather than by telephone. Fewer than half the wives attempted to use the telephone on the first night, and of those who did, several reported that they could not reach anyone by telephone, as the lines were jammed. Ten of the 17 wives of trapped miners did not go near the pit head on the first evening. This is in marked contrast to the less involved wives. The great majority of wives of both off-shift miners and personal service personnel went to the pit head. In the Minetown disaster, personal convergence (cf., Fritz & Mathewson, 1957) varied inversely with the degree of personal involvement in the disaster. Most of the people who went to the pit head acted as helpers.

All the trapped men were working at roughly the same place in the mine. Their wives knew this, yet on the first evening their estimates of when they might expect news of their husbands varied considerably. Eleven wives of trapped miners expected news of their husbands immediately; 4 expected news within five hours, 2 wives did not expect word for days. The four wives who had husbands in the 1956 explosion had the least immediate expectations. Those women, then, who had been directly involved in an earlier disaster had a different set of expectations from those who had not had this experience.

During the first evening, most of the wives in the three samples visited the homes of friends or relatives, or had visitors call. The implications of the disaster were community-wide, and there was a great deal of visiting going on all over Minetown. Almost all the wives of trapped miners stayed up all night, most of them with relatives Despite the general indication in social science research of the declining importance of the extended family, the extended family in Minetown played an important role during the crisis and in the later recovery from disaster. The importance of satisfying the need to be with others and of maintaining a stable, supporting, interpersonal environment stressed by Fritz and Rayner (1958) is well illustrated by the behavior of wives during the night of the bump.

The Waiting Period

Table 2 shows the frequencies of a number of behaviors in the three samples of wives during the waiting period. During the long period of uncertainty while waiting for news of their husbands, the wives were living within a relatively stable, intact, social structure. Three wives moved into the homes of relatives. Fourteen of the 17 had relatives staying with them night and day, and they kept their houses operating very much as usual, preparing meals and accomplishing the daily routine with the active assistance of relatives. This routine extended to laundry and baking. Only 3 wives stayed alone any night.

Unlike many disaster situations, there was no major widespread devastation or reconstruction. This made it possible to maintain social groupings, especially family life with a routine but meaningful activity linking it with ongoing society (cf., Tyhurst, 1954a). The routine was retained, assisting the period of waiting, but most wives did not have sole responsibility for maintaining it.

Whenever a district bump occurred, it was taken for granted that there was no school the next day for children whose fathers

TABLE 2

Three Groups of Wives Having Different Levels of Involvement:
Waiting Period Response by Per Cent of Group

Response	Wives of trapped miners (N = 17)	Wives of off-shift miners (N = 11)	Wives of professional and service personnel (N = 7)	All wives (N = 35)
Stayed at others' homes	27	27	0	22
Had others to stay with them	83	9	0	43
Prepared usual meals in own home	83	91	100	89
Received help in meal-making	83	0	0	40
Did laundry	71	64	71	69
Did baking	59	73	57	63
Kept children home from school till twelfth day[a]	83[b]	13[c]	0[d]	52[e]
Went visiting	53	73	71	63

[a] Appropriate only for wives with school-age children.

[b] N = 12.

[c] N = 7.

[d] N = 2.

[e] N = 21.

were involved. Two wives sent their children back to school before word was received of the fate of their husbands. Ten wives with school-age children waited until the rescue before returning their children to the school routine, thus keeping the family intact during the waiting period.

It has been found that adequate and authoritative information, widely disseminated to the public, will discourage the speed of disruptive rumors and combat reactions of either unwarranted defeatism or unwarranted optimism (National Opinion Research Center, 1950). Press, radio, and television coverage was extensive, and originated from a building at the pit head. Periodic announcements of progress were given. The majority of wives continually listened to the broadcasts of a nearby radio station which for the first few days played only recorded music between news flashes and commentary on the disaster. Nearly half the wives read all the newspapers they could obtain, "mostly for interest only," rather than for factual information. The radio seemed to be their most reliable source of news, as it was in the West Frankfurt mine disaster (Conference..., 1953). Some wives even watched their rescued husbands removed from the mine over television and then drove to meet them at the hospital.

Despite adequate information, rumors develop in disaster (Caplow, 1947). Six wives said that they did not hear any rumors, and 5 said that they heard rumors, but did not believe them. Several wives maintained their private source of news, which for them was dependable: they relied on "official information from my family doctor" or "firsthand information from my brother who was a dreagerman."

On the fifth day, a high ranking official made the public announcement that the company held little hope that any man remained alive. Although an official announcement, it was rejected as untrue by over half the waiting wives. The majority had more faith in their own or others' experiences (and hopes) in these matters than they did in the official report "My brother-in-law is a dreagerman and he told me they could still be alive." Of the total, 11 said they "had hope for the whole period and never gave up." For the other 6, however, this was a turning point from hope to hopelessness: "I lost hope. It was an official announcement, not just a rumor," or, "I believed it. I went out and bought a black dress and hat and told the children."

Two wives were erroneously advised of the death of their husbands by telephone without notification by their clergyman. Although the official role of the clergyman as the official bearer of death notification could and did remove much error and neutralize rumor and false reports, as well as giving specialized support at this time of crisis, it restricted the clergyman's other activities. Two wives who were visited by their clergy in their role as clergy, rather than as official notifier, admitted being frightened by the visit. Several churches organized visiting committees of church members rather than bring additional stress to the waiting wives. Nevertheless, over half of the waiting wives were visited by their clergyman on the fifth day, the day of the official public announcement that little hope was held for any of the men.

When asked what was the worst thing about the whole waiting period, 9 wives discussed the inactivity and sense of insecurity of waiting for days "without knowing what to do." Most reported that they felt helpless and had depended completely on radio and television. Only a few had any difficulty in sorting out the days in retrospect. Most of the wives said that they had not slept much for the whole waiting period.

Two wives felt that the visit of the clergy distressed them, especially those whose prayers "took it for granted that my husband was dead." Two complained that the radio station continually played inappropriate music between announcements, especially Good Night, Sweetheart. Two complained bitterly about the official company announcement on the fifth day that caused so many of them to sink into a mood of hopelessness. The two wives who had been adversely affected by rumor said that this had been the worst part of it. One wife had a daughter that became hysterical, and another found that when she cried "the baby also cried," so after the first day she was determined not to cry.

Several family patterns indirectly aided the wives to maintain themselves and their household during the waiting period. Thirteen handled the family budget, and 10 had a weekly grocery account. Directly applicable to the disaster was the fact that 13 of the 17 wives had discussed with their husbands what to do in their day-to-day living in the event of an accident. Only one was unable to carry out the proposed plan, as she had just given birth. Most wives stated that their husbands had warned, "don't go near the mine."

Some wives were better prepared for disaster than others in that they had credit accounts or in that they customarily handled the family money. Some families were better prepared in that they had discussed plans for day-to-day living. By these indices, 6 of the wives were fully prepared. One was totally unprepared and 10 were prepared to some extent. Three of the fully prepared wives had had husbands trapped in the 1956 explosion.

Conclusions

The consideration of the rescue operations, the surface relief support, and the waiting wives leads to three conclusions.

First, the rescue operations could have continued in the way they did only with the support of a large, stable, social structure behind them. All services in the community and area remained intact during the crisis.

Second, the community itself and many formal organizations were prepared to cope with this type of disaster by virtue of previous disaster experience. Not only relief agencies but the miners and citizens themselves had been through a major disaster two years before. Many of the same patterns were followed.

Third, the social structure and institutions of the town were able to meet this particular type of emergency. The evidence at hand does not bear out findings of other studies that the disaster had little direct impact on the lives of the inhabitants (Gordon & Raymond, 1952), or that the people were complacent, apathetic (Gordon & Raymond, 1952), or fatalistic (Luntz, 1958, Marks & Fritz, 1954). Minetown was not similar to Coaltown, which was characterized by impersonality, resignation, cynicism, and suspicion (Luntz, 1958). Rather, over many years, living from day to day with dangers, the population acquired appropriate values, beliefs, and social organizations. These social developments, although differing greatly from those in communities that do not face ever-present danger, were similar to those of other mining communities.

As in many coal communities, workers entered mining because, from their perspective, mining was the best of the few alternatives they saw available. In Minetown, it was the highest paid industry in the area. Once an experienced miner, a man had profitable but nontransferable skills and a fierce pride in craft; at this point, mining was spoken of as "in the blood."

In the face of a constant and uncontrollable danger, the miner, like countless generations before him, controlled his relationship to danger through shared myths and beliefs (cf., Malinowski, 1925). In Minetown, these included unlucky days--Thursday was the current one, although Friday had been the previous choice. Absenteeism was a way that the miners coped with the danger involved in their work situation. While at work, however, the joke was paramount, not the discussion of danger.

Besides shared myths, many miners worked out their own intuitive signs as a guide to impending danger, which allowed them to continue working in the mine. These private guides to danger included the action of rats, sounds heard in the mine, or tensions felt. "I heard the props and stringers cracking," "the pavement would get hard," or "I feared her when she was quiet."

In common with many mining communities, the norms shared by all individuals guaranteed mutual help. The miners' code of rescue meant that each trapped miner had the knowledge that he would never be buried alive if it were humanly possible for his friends to reach him. This code was so widely understood and so unconsciously accepted that no miner-rescuer was faced with serious role conflict. At the same time, the code was not rigid enough to ostracize those who could not face the rescue role.

These common expectations, norms, and social arrangements extended to the families of the miners. No family in the community was untouched by the disaster, differences existed merely in degree of involvement. The majority of families had plans for an emergency. Their financial arrangements permitted a wife to look after her family during a period of waiting and immediate crisis. Those directly involved or bereaved were assisted by the extended family as well as by those less directly involved.

This level of preparedness, which characterized the social structure of the town, extended to the growth of particular institutions, as well as modifications of traditional institutions. A disaster fund had been in continuous existence from 1886. Besides the union fees and employees' relief fund check-off by the paymaster of the company, there was also a doctor's check-off and a church contribution check-off. The clergy traditionally bore news of a fatal accident to the family on behalf of the company.

Unlike communities infrequently and unexpectedly visited by disaster, Minetown was constantly prepared for this anticipated eventuality. Quite unconsciously, through constant dealing with danger and death, patterns of behavior became established and, through time, so widely used that they can be considered norms, codes, popular myths, and institutional arrangements. This appropriate social structure allowed for, and cushioned, many social effects of disaster.

A mine disaster or emergency occurring within such social arrangements is not likely to arouse either apathy or blame and indignation. To the outside observer there may even be an air of normality. This normality could well have been anticipated, for Minetowners knew what to do; their behavior had been patterned over many years.

CHAPTER 4

BEHAVIOR OF THE TRAPPED MINERS
A DESCRIPTIVE ACCOUNT

When the bump occurred on the evening of October 23, 174 men were underground in No. 2 mine. Within 24 hours, 81 were rescued, one of whom subsequently died in a hospital. Of the 93 others, 74 died in the mine and their bodies were eventually recovered. The remaining 19 were trapped and rescued alive 12 in one group, 6 in another group, and one in semi-isolation. The group of twelve was rescued after 6 1/2 days and the group of six together with the semi-isolated miner after 8 1/2 days. This chapter describes the behavior and experiences of these 19 men.

Situational Factors

The Group of Twelve

The miners in this group ranged from 22 to 58 years in age, with an average of 38.8 years. They had attained an average educational level of grade 6.9, with a range from grade 2 to grade 11. Only one miner had gone to high school. All of the miners were married and living with their wives, except one man who had been separated from his wife for a considerable number of years. They had an average of 2.2 children. Six of the men were contract miners and 6 were datal workers. The members of the group had spent from 3 1/2 to 31 years in the mine, their average was 18.6 years. The fathers of 9 of the 12 had also worked in the mine.

The area occupied by the trapped group of twelve was at the bottom of the 13,000 foot level wall. Ten of the twelve remained together in a cavity measuring approximately 12 by 30 feet (see Figure 3). The two miners who were immobile with severe injuries were in an adjoining cavity of about the same over-all dimensions. The roof in the second cavity was more dangerous, and there was less clear floor space. There was a third, small cavity about 8 feet from the location of the ten miners, it was entered by a hole "large enough for a man to crawl through." The miners were vague about its dimensions, but it was established as being approximately 2 by 10 feet with height ranging from 3 to 4 feet. The ceiling was described as ragged and dangerous. This was the cavity that

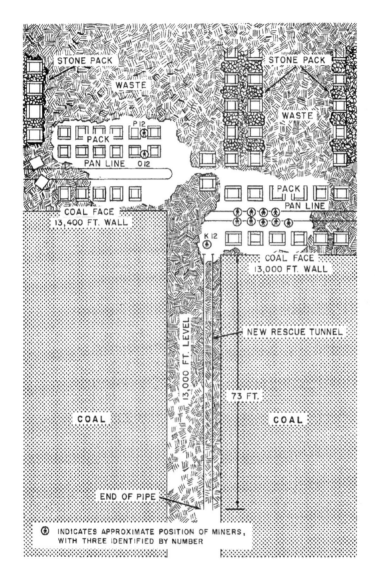

Figure 3. Location of the Group of Twelve While Trapped: Ground Plan

contained the air pipe through which contact was made with the rescuers. It is notable that the trapped miners gave the dimensions of the cavity where the ten were located as about 10 by 11 feet, one-third of what it actually measured.

The main tunnel used in the escape operations was described as very narrow in parts, "only big enough to crawl along on hands and knees." It was about 150 to 200 feet long, and its width and height varied. There were many obstructions in it packs, broken pipes, and dismembered dead bodies. A few shallow cul-de-sacs led off this tunnel, all of which were explored by members of the group.

The Group of Six and the Semi-isolated Man

The ages of this group ranged from 29 to 49 years, with an average of 39 years. The average educational level was grade 6.6, ranging from grade 4 to grade 11. Only one had attended high school. But for one bachelor, all were married and living with their wives. They had an average of 3.5 children--one of this group had 12 children. Four of the men were contract miners and 3 were datal workers. Their average length of time in the mine was 15.8 years, ranging from 9 to 32 years. The fathers of 5 of the men had also been miners.

The group of six was trapped at the top of the 13,000 foot level wall (see Figure 4). The dimensions of the main cavity were 30 to 40 feet long, about 10 feet wide, and, due to a sloping floor, from 4 to 6 feet high. The usable space was cut down by two packs and a broken engine. There were also pans and an air pipe which ran through the cavity at ceiling level. The ceiling of the cavity was ragged and dangerous in places.

The man in semi-isolation (X6)* was located in a small cavity about 75 feet away from the group of six. This smaller cavity was about 4 feet wide by 4 to 5 feet long and 5 to 6 feet in height. A small hole about 2 to 3 feet in diameter led into this cavity from the main tunnel that was subsequently used in escape attempts. This opening allowed air to enter the cavity, permitting the survival of X6.

The 75-foot tunnel used in escape operations was very narrow and was described as hardly big enough in parts "for a man to crawl

---

*Each member was given a code name the letter identifies the individual while the number identifies the group to which he belonged.

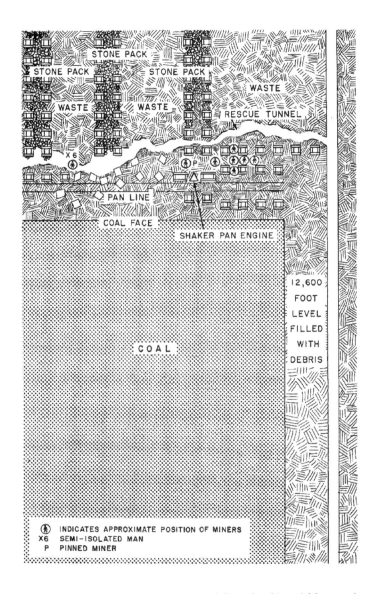

Figure 4. Location of the Group of Six, the Pinned Man, and the Semi-isolated Man While Trapped: Ground Plan

through on his hands and knees." The roof was dangerous and there were shallow cul-de-sacs leading into the waste.

Physical Dangers

The temperature in the cavities containing the two groups was approximately 80° Fahrenheit, and the air was stagnant. The air flowing from the depths of the mine to the surface in the days following the bump had a high concentration (about 70 per cent) of methane and carbon dioxide. There are no exact figures available on the gas concentration in the cavities occupied by the trapped miners.

Regulations required the mine to be evacuated when the methane content had reached 2.5 per cent. During the last day of rescue, a calculated risk was taken in sending barefaced miners into the return airway of the 12,600 foot level in a concentration of about 6 per cent methane. Several of the trapped men said they found breathing heavy at times. One said, "I had an awful hard time to breathe. I think it was right after the bump and I think there was quite a bit of percentage of gas." Another said the gas only bothered them for about ten minutes after the bump, after that, the air was quite good and his breathing was normal. But there were pockets of gas at high concentration trapped in some of the passages through which the men attempted to escape, and a few men lost consciousness and had to be pulled away by their companions.

The anaesthetizing threshold of carbon dioxide is considerably lower than that of methane, and the prolonged unconsciousness of the semi-isolated man in a restricted space may have been due to a high carbon dioxide concentration. His survival without food or water with apparently few mental or physical effects resembles the state of "hibernation" induced by an 11 to 17 per cent carbon dioxide concentration in experimental animals (Seevers, 1944).

The narcotic effects of carbon dioxide at a concentration as low as 5 per cent may have had something to do with the relative freedom from overt anxiety and panic of all the trapped men and the freedom from pain of those who were slightly injured (cf., Meduna, 1950, Stokes, Chapman, & Smith, 1948; Wolpe, 1958). The carbon dioxide may also have contributed to the hallucinations which most of the trapped men reported (cf., Meduna, 1950).

The members of both groups were aware of the hazards of gas. In both groups there were men who had been in the 1956 explosion and who warned the others immediately after the bump to keep their heads near the ground as a precautionary measure.

Once the impact period was over, however, fear of gas did not cause undue anxiety except when they explored new passageways.

The danger of another bump was always present, but none of the trapped men spoke about it while underground. They did worry, however, about the possibility of the collapse of the ragged roofs of the tunnels and cavities, particularly when they were making escape attempts.

Psychological Dangers

The fear of not being rescued was uppermost in the minds of the trapped miners. The sounds of the rescue operations were interpreted by some as the sealing off of the mine. This fear was dealt with by denial and by their companions' more hopeful, reasoned explanation.

In some miners, fear changed to resignation during the last few days when the sounds of rescue occasionally stopped for an hour or two at a time as the rescuers in the long, narrow tunnels changed shifts. These were the periods when most anxiety and pessimism were expressed. In the group of six, underground for two days more than the group of twelve and without contact with the rescuers until the final breakthrough on the ninth day, there was more ambivalence about the interpretation of the noises of rescue.

The group of six had two additional problems: the pinned miner and the semi-isolated miner. The presence of a pinned and dying miner whom they were powerless to help was a serious threat to the morale of this group, particularly because this man lived 5 1/2 days. When the bump occurred, he had been working close to a pack. One of his arms was caught between two of the crushed timbers of the pack. His whole arm was pinned down and crushed below the shoulder joint.

D6 described the situation surrounding this injured man as follows:

> One fellow was caught. He said he couldn't get his arm out...Well, we looked around to see what we could do, and we had no saw and no tools at that time. And we thought about trying to comfort this man as best we could...We gave this man some aspirins...We shared our bit of water with him...He had his arm crushed right down solid. The whole packs, in between. Yes, that was squeezed in about half an inch, down solid.

In the 2 1/2 days during which the pinned man remained conscious, he moaned a great deal and pleaded with the other trapped men to remove his arm. One miner reported that the pinned miner said to him, "Oh dear, oh dear, E6, I would help you if you was caught." However, they decided not to remove his arm. D6 explained it this way:

> If we could have released his arm it would have been a terrible painful pressure to have released his arm...the way the man was, would have become violent and could have died...and the pack was not too stable anyway. We would have had to have climbed on the tops and sides. I suppose we could have got the man out with much, much,--it would have been a terrible struggle trying to get him out without hurting him ...I knew that once his arm was off and the flow of blood started...we would have to put a tourniquet on ...and...release it so often, and when we released it we would have weakened that man at the same time. What I mean you have to let a certain amount of blood come, and that man will probably become violent... And we thought the men would probably be within two or three days up to rescue us...The boys all agreed with me.

E6 put the problem in these terms:

> (The pinned man) wanted us to cut, to cut his arm off Well, we couldn't very well get a saw in there, well, we tried with a pack chisel but it was too dangerous for the whole bunch of us...we thought we would start the blood and then we would have trouble with him, maybe go out of his mind, see, and then if anything did happen, well, the way I looked at it, I thought if anything did happen and we cut his arm off and then he died that way, they'd kind of blame it on us.

During the last three days of this man's life he was delirious. While in this state, he continued to plead with the group to amputate his arm and raved about water much of the time. He died quietly on the fifth day.

Guilt and ambivalence were also noted in relation to X6, the man in isolation. When first contacted on the second day, X6 was thought to be "pretty near gone." On the second contact with X6 on the fourth or fifth day, they discovered he was still alive and had moved; however, they were very tired, weak, and without light, so they decided not to move him up with them.

Impulsive and irrational behavior was occasionally a threat to the trapped miners. In the group of six, F6 was near the breaking point at times. He was inactive, pessimistic, and extremely irritable. On one of the last days he said, "If I had a knife I would put it through me now." He was more bothered by the pleading of the pinned man and once had to be restrained by C6 from impulsively severing the pinned man's arm with an axe. In the group of twelve, M12 had an outburst of hostility and on the seventh day became very "riled up" when he thought the rescuers were digging in the wrong direction. J12 was inactive and tearful. The two injured men expressed resentment and suspicion about the rationing of water. Such situations were handled by the other miners by direct action or reason and reassurance.

The presence of dead and decomposing bodies was a source of stress for both groups. I12 said,

> It really shook me when I looked around that pack and saw a man thrown to the roof, and said, "My God! look in there," but it is best not look at all...Seeing his head, the way it was battered, it took the life out of me for a while.

C6 said about the smell of the decomposing bodies,

> Oh, it was terrible the last two days. It wasn't too bad until--see, the place was so small--we didn't --we had to stay in the air to live ourselves. We had to stay in the air because if we had got in the waste a little bit she was dirty, gassy. So we had to stay right in the air...And this fellow that died at the engine started to smell. We turned him over. We couldn't move him all together. But we turned him over and got an old piece of canvas and laid on him to try to get the air to go up by us that way. Oh, it was terrible the last couple of days. The smell, that is what I was scared of. The stuff might poison us, you know. The fume of that. I was scared of that too. Of course, you can't tell yet what it might do to us.

Few of the men spontaneously mentioned the dead bodies, and some claimed that they were unaffected.

Behavior While Trapped

M12 described the bump to the interviewer:

This young fellow...he come down; shifted from his place above to down below. He came down to me and just made a wisecrack like we do all the time. We are all the time saying things. "Are you going to load the pans or dig some more?" I walked over to him and I said, "Boy, I'm not going to take that stuff off of you!" And I started over towards him. And he was setting on the pans. And I said, "Oh, by the way, did you feel that bump a while ago?" And just as I said "bump," that is when the big one come. It just seemed like the word <u>bump</u> was the trigger. Immediately I said that, why, he flew. I seen him going I seen his feet, that was all. I thought at that time--I thought that the roof had come in on us, that we were up to the roof, right up, and it seemed like it expanded the way I went. I remember going right to the roof and I thought I was killed. Immediately after that it seemed just a matter of seconds, it happened so fast and expanded. But I could see it all, the way I looked at it, just like a bullet shot out of a gun. And then I could hear the boys moaning and groaning.

There was just one fellow that was caught bad, and one of the young fellows was working on him, and that is all the room he had--just for one fellow. It didn't take him very long, and he got him out. Well, then we started looking around for each fellow, to hear and see how he was. Well, after a while we got all together and we found out there was none of us seriously hurt. We weren't seriously hurt but we heard another fellow down below us, hollering. So we went down to him, and it was this P12. By just looking at him I knew his leg was broke and he was in a bad way. So we told him... "You are safe as long as you are setting where you are." We didn't think at that time that the roof would come in on us. So we came back up to this little place where we were all setting there. We were there for six days.

Of the twelve men working at the bottom of the 13,000 foot level who survived the bump, two were seriously injured. O12 suffered from an inferior dislocation of his left shoulder with damage to the brachial plexus. He also had several fractured ribs on the left side. Both were very painful conditions that prevented much movement. P12's injury at the time of the bump resulted in a comminuted fracture of his right femur. He also had multiple contusions of both legs. These injuries rendered P12 practically helpless.

None of the group of six was seriously injured by the bump. F6 was buried to the waist in coal and extracted by E6. His hips were very painful for two days and he did not take an active part in any of the group activities. Immediately following the bump, the group clustered around C6 who had been trapped for four days in 1956. He advised them "to stay together" and to "keep their heads down" because of the danger of gas. They remained thus for a few minutes and then B6 and D6 went to the assistance of the miner whose arm was crushed. Then they all sat down and talked about what they could do.

The Escape Period The Group of Twelve

M12 gave this account of the escape attempts.

> We went back up there (after pulling men from the rubble) and sat down and started talking about it amongst ourselves. And right at that time, I being the oldest miner and more experienced than any, I told them I never seen a bump yet that there wasn't one way out. There was always one way, and we figured this out amongst ourselves and we were all going to get together and work...One of the other miners and myself, we talked it over and we said, "Now the air is coming to us very good, so let's try and work our way up. As long as we can work our way up, the air ought to blow the gases away from us." So we started to work our way up the wall, the 13,000 (foot level) wall. Well, we worked our way up there and we got up three stone walls. We dug through three stone walls and we come to the next part of the waste where the stone wall was above a bit and we seen that the only way we could possibly do it was to go right up over the top of the waste. There was not too much space between there but we knew that that was our way. So we started up through there and just as I landed at the top of the waste I run right into gas. Just the minute I hit it--I had more experience--I knew what I was into and I immediately wheeled around and threw myself back so I could get

out of it. I knew I was into it, but I got back out again and afterwards I sat there and I said to myself, "Now, did I really run into gas or was I scared?" I had a young fellow with me. I wouldn't say that I didn't have any faith in him, but I would say I would have had more faith had I another experienced miner. So I set there and I thought to myself. was I scared or did I really run into it? So I thought there for a while. I had to really convince myself...So I said, "Boys, keep right close." And I called one of the other boys up and I said, "I want you next to me; I am going to try it again." ...I guess I wasn't scared because it knocked me down again. I figured that was enough right there.

So we decided then we was going to try down (the wall). So we started down and this P12 (a man with a broken leg) said, "Don't leave me here. Get me up somewhere where I feel safer." So O12 was laying up there with a dislocated shoulder so I told two of the boys, I said, "Now the two of youse go up to where O12 is. We will go down and start working on down below." So they took hold of him (P12) and took him up. They must have took him fifty feet, for all he had a broken leg and everything he never said a word though; he stuck right with them. Well, we went down and started to dig through the stone wall there. We got down through the first one and we got to the waste from the first one ...and it was so ragged and so bad it disturbed the boys and they decided they didn't like to take a chance on it. So we all talked it over there and decided to go back. So we went up again and sat there and talked.

After about three days of exploring every possible escape route, the eight active members of the group of twelve were exhausted. Their lamp batteries were almost dead and the water supply was depleted. They decided that escape was impossible and returned to the cavity where they were found on rescue.

The two injured men lay in their cavity alone during the escape attempts. They had known each other previously. They decided to conserve their energy and put all their hopes on rescue. O12 strapped the arm of his dislocated shoulder to his side. He did get up and make one unsuccessful search for water.

There was not a major morale problem during the first three days in either group as the active members occupied themselves most of the time with exploring possible escape routes and searching for water. "The first three days went by just like--well, I didn't realize it was Monday." They were tired following these efforts, which at times involved hacking their way through stone barriers. During rest periods, the conversation centered mainly around chances of escape or rescue, and occasional discussions about the welfare of their families.

The Escape Period: The Group of Six

After describing the confusion of the impact, E6 said,

> Then I see what happened. Well, I thought, it's just a miracle it didn't happen here. And I said, "Maybe down below it's pretty good, if we can get down to it." Well, the shortest way, we thought, it was to get out the upper way. We thought we'd go up the pack and get out. Well, we tried and couldn't. We came back and we tried...over the waste. We went down and looked around there. We had lights then. And we looked around and we seen X6 (the semi-isolated man) on the left side. Well, we thought he was gone, pretty near gone...Well, he was unconscious, you know. Just wasn't himself. We seen where he was trying to get a little air through so we opened it all up and we got air through.

C6 discussed later escape attempts:

> Well, the first 24 hours after we got first settled down there, we found an old saw and we tried to escape. Well, the escape was only about that high I guess, about five inches. Well, we dug a hole out through the waste and out through the top of the escape, and we sawed two booms out--two old wooden booms--we sawed them out. The old saw wasn't much good though. The handle was broke off of it. Well, then we seen we couldn't get out there, and we cut up through the high side...Then we come to the stone right up like that. I said to the boys, "If we start that stone now, we've got no lights." The lights started going dim, see. We were only using one light between us at the time. We went out three different times to try to get up the high side, that is, the escape of the 13,600 (foot level). And we seen we couldn't do that. I said, "Boys, the best thing we

can do--our lights is gone out-- is to go back to the
wall and lay down and stay there. Our only chance
is if the men gets to us in time."

The Survival Period. The Group of Twelve

On about the third day after the bump, the group of twelve,
without water, food, or light, terminated their escape attemps.
M12 said,

> I just laid back and I said to myself, well, I guess
> it is up to a higher power than us to get us out of here.
> I said the only thing I could do, I figured, was to lay
> back and just try to keep our minds occupied with every-
> thing else but what was going to happen.

They had had a few sandwiches, a doughnut, and a few pieces
of chewing gum among them, but no attempt was made to ration this
small quantity of food. A few men were able to find unsmashed
water flasks, either their own or those of dead men. They did not
begin to ration water until about two days after the bump, when
much of it had already been drunk. What remained lasted only until the morning of the third day. Another gallon of water was found
on one of the last escape attempts (probably on the third day), but
it too was soon gone.

M12 said,

> We never thought about the water until it was too
> late. We never thought about water or grub. We
> missed out on that, see. We missed the boat there...
> We could have rationed that water (sooner) and that
> grub. We could have had water, with the water we
> did have at first. If we had done what we done after
> the third day, we would have had water. Not lots of
> it, but we would have had enough to do us through
> them six days. But we never thought about it.

Q12 told the interviewer,

> The lights were done, no water or nothing. So
> ...K12 said, "Let's drink our own urine." So he
> tried it. It was right salty. Oh, you could hardly
> drink it. So I said, "Boys, we have to survive.
> That is what we will have to do: drink it if it makes
> us sick or not."

Seven of the group of twelve succeeded in drinking urine. As in the group of six, they adapted themselves to it by stages. They first wet their lips, then rinsed out their mouths, and then drank. The time required for the adaptation varied from an hour to two days.

Five of the group of twelve did not drink urine. Three did not try. One reported: "I wasn't thirsty. I would have if it had gone down to the worst." Two tried and failed: "I tried to drink it but vomited it up, and I didn't try again."

Coal and bark were put in the urine to try to improve the taste. Some chewed bark to try to extract moisture from it.

Some of the men who had been active during escape lost hope when escape proved impossible. M12 told the interviewer,

> I really gave up. I thought to myself--after what we had tried, the conditions I had seen, all the walls in--I can't get out and I am just as good an experienced man, how are they going to get in? So I just told myself it was all over...I thought that if those fellows outside that was trying to get in to us, I figured they would say there was nobody alive, there just can't be. So instead of digging a hole to get in to us, that they would timber the place all up, getting ready to bring bodies out. And I figured they would never get to us that way. It would be weeks and weeks and I said I know that we can't survive that.
>
> We could hear the boys when they were digging and pounding for us. We could hear them way off. Oh, it sounded like an awful long way off...I was listening to the different sounds they were making and it seemed to me like I could hear them sawing off props and it seemed like I could hear them pounding props and I knew. I was just piecing the thing together and I was trying to picture what they were doing and that was the only thing in my mind--that they were putting a boom up and sawing off these props and pounding them in, getting them in, see, by doing that, that was the whole level, so they could get big boxes in and out. And that was what was going through my mind. Never dreamt in the world they were digging a small hole... I couldn't picture them doing that.

At one stage, M12 threatened to "put his fist through" a man who made a noise while they were listening for the sound of rescuers. M12's despondency and irritability were shared by others. J12 was pessimistic most of the time and contributed little to the group. H12 retired behind a "schizoid screen" and had little hope of rescue. The two injured men remained emotionally isolated from the others. According to O12,

> The last day (before contacting the rescuers) we were all saying good-bye to one another as we thought we were dying, we would not see one another again.

But others did not give up hope. Q12 said to the interviewer,

> Then we could hear them working for a while outside. But all the time this was going on we could hear pounding on both sides of us. We thought it was up the wall, we thought it was down the level or in the other wall, and then it sounded like it was at our level. It sounded like miles away. And we pounded on pipes and we pounded on the pans. So then we crawled up the hole where the air line was. It had a big shut-off on the valve. It was reduced, I think, from five inch to a three, and where the three came on to the five it was broke off...So I felt around and I found this pack wedge. And I crawled out in the hole, working in the dark, started chiselling at this here nut on the shut-off valve. I got it off. And I tried to turn the couple around...I got one loose and I turned it around and tried to get it off but I couldn't. I hollered at K12. I said, "I can't stay here much longer, somebody had better come in." So I went in back of the hole...At last he got it pulled off. Well, then that gave us a full six-inch air line, you see. So we listened there and listened there. We heard a motor running out there...and it was running fast. We knew they was working hard. You could tell they were getting further in because it was running longer each trip. So we waited, waited, and waited.

Six other men took turns listening and shouting through the pipe, but K12 and Q12 played the major role. Q12's description continued.

> We laid there side by side. My face was on the bottom and his face was on top of mine with our noses stuck in the pipe, listening. We listened there for

pretty near a whole day...So all at once K12 heard
somebody's voice speak out there. Well, he hollered.
Then he got an answer back.

The first contact of the group of twelve with the rescuers was made late on the sixth day of entrapment. The rescuers were separated by 73 feet of solid rock. Through the pipe K12 yelled the names of the twelve trapped men, that two of them were seriously injured, and that they required water urgently. The rescuers broke through twelve hours later, having previously supplied them with water, tea, and soup through a smaller pipe which was passed down the main air pipe. On realizing that they were safe most of the men wept for a few minutes. A doctor accompanied the rescuers into the cavity and remained until the last man went to the surface.

The Survival Period   The Group of Six

Late on the second day of their entrapment, the group of six heard sounds of rescuers. D6 said,

So we figured they had cleaned up our level there, and you could hear the rumbling, something like the boxes rumbling. Sometimes you could hear them and sometimes you would never hear nothing for four or five hours...I said, "The thing is we have got to get out messages, got to let those fellows know we are alive."

Shifts were organized to beat against the pans and pipes in the hope that the rescuers would hear them.

D6 continued:

The worst thing I found was these rescuers. There were times, there were times I said to the boys, "We'll wait when they are changing shifts and then we will start banging and pounding." And we used to holler-- for help, you know, all the voices. I said all our voices would carry better over the rubble and stuff. I figured, well, we can hang on for a few days at least...And Saturday came along and I said, "Well there is one thing, boys. I want to be up now. You never can tell, they may not work over the weekend, you know...We want to pound on those pans every hour on the hour, right off the bat."

Although the group rationed their water at D6's suggestion, when their hope of immediate escape or rescue had faded, the water was finished by the second or third day. E6 told the interviewer, "Our water was all gone, so I said, 'Well, boys, we're going to have to drink pure pee.' And they kind of laughed."

They began to consider E6's suggestion seriously within a short time, and, after some gagging and vomiting, all drank urine. "One fellow was drinking it all the time. I thought he was drinking beer." A single chocolate bar was found by the group of six and they broke it into six pieces and ate it. This made some of them more thirsty.

A6 said, "At one time I thought about cutting my leg and getting some blood to drink." F6 annoyed his companions considerably by talking about "pop" during the last five or six days. He said,

> I thought a lot about dying. I thought dying of thirst would be a horrible way to go. My tongue was so thick, my throat so dry, I couldn't swallow and could hardly talk. It wasn't pain. It was a terrible despair for water. If someone had come up to me with a gallon of it and said, "Drink this and die in ten minutes," I'd have drained it down and wouldn't have cared.

E6 described his faith in rescue when the escape attempts had failed

> So we set down and pounded on the pans and wondered what we'd do and kind of slept like. And then all at once, oh...F6 said... "You still got hopes?" And I said, "Yes I've got hopes...I got good hopes and I will until I draw my last breath."

D6 expressed similar faith in rescue:

> I couldn't understand us living through and living that long (unless rescue was to come). We were in fairly good condition. Nobody went haywire. We were all talking sensible, ordinary conversations. And I figured if we were left that long in fairly good shape, we might have been a little on the weak side but no one was delirious or unconscious, we were allowed to keep good.

Not all of the group of six claimed to have this faith in rescue. C6 said to the interviewer,

> I sat there and thought, "Now I wonder how long?" Of course, I just said to myself, "Well, I'll likely not know much about it because I'll keep getting weaker all the time." The way I figured it we would just keep getting weaker all the time, and I didn't think we was going to get out because I didn't think they were going to get to us. I figured we would just lay there when we got that weak. Well, in fact, the last day we couldn't hardly stand up.

C6 had been active immediately after the bump, but during the survival period he cooperated only in a passive manner. A6 was pessimistic and helped only when asked to do so by D6. F6 became irritable, depressed, and uncooperative.

D6 described their rescue on the ninth day.

> I had just finished pounding, and one of our boys--that was their turn--they got up and started pounding on it. And I laid back and kind of dozed off. And then we heard, "How many is there?" We never heard those fellows until they come right up to us. We never heard anything...I woke. And one of the boys grabbed me and kissed me. And we said, "Thank God you are here" or something like that, you know. And then the rescuers, he and I, joked back and forth and carried on. And the first thing we hollered for was water.

The Survival Period: The Semi-isolated Man

X6 was trapped for 8 1/2 days about 75 feet away from the group of six. He was without food and water during the entire period. He was contacted three times, several days apart, by members of the group of six. X6 had only vague memories of the trapped period. He said, "It all seems like a dream." He is considered in more detail in Chapter 8.

Leadership

In this section, the term _leadership_ is applied to those miners who played a directive role with reference to critical problems or events. The evidence for such a role was taken from the man's own interview if his statements were confirmed or at least not contradicted by another miner's account. The critical issues for the trapped miners included

   (a) being partially buried and in danger from gas (the immediate post-impact situation),

   (b) escape,

   (c) rationing of water,

   (d) drinking urine,

   (e) maintaining group morale,

   (f) getting a message to the rescuers.

The group of six faced one other distinct critical problem. what to do about the pinned man. Problems (a) and (b) were concentrated in the escape period, while the others applied mainly to the survival period.

The following accounts for the two groups focus on the miners in each group who played central leadership roles. A comparative and quantitative analysis of leadership is presented in Chapter 5, and the personal qualities of leaders and their roles are discussed in Chapters 7 and 8.

The group of twelve. A senior miner, G12 behaved most adequately right after the bump. While others were dazed and confused, this man helped those who had been partially buried. K12 said, "In the meantime, G12 has come up to where N12 and me was and uncovered him (N12) ..and of course G12 went by me to help the other boys out." G12's behavior on impact was direct and appropriate, and the other members of the group seemed to expect this of him. During the escape activities he participated with the others but did not take a leading role. Nevertheless, the group seemed to expect him to take responsibility throughout the period of entrapment. He played a part in rationing the water, J12 said, "That was a good thing because if you passed the can around, some of them couldn't resist taking a good big swig." On the seventh day, after they had contacted the rescuers and received coffee, water, and soup, M12 became very disturbed, thinking the rescuers were digging in the wrong direction. K12 told the interviewer

   I said, "Now look, M12, I'm not a miner, but I could tell why they were close."...I'm what they call

a shiftman. So I said, "We'll get G12," This M12
was the chap that said this to me, "G12's an old
miner." I said, "G12, you crawl in there and give
them directions."...G12 had been a miner and he
was a good one.

G12, however, apparently did not contribute much to the
general morale of the group, though he was looked to on several
occasions. He had received psychiatric treatment for a depressive
reaction two years earlier.

One of the most active miners in escape attempts in the group
of twelve was M12. Although he got angry the first night and was
one of the most pessimistic during the survival period, he was foremost in promoting escape. K12 said, "He'd drive us after he was
down on his back...to do some job to try to get out."

K12 was clearly the leader in the survival period, although
he acknowledged the seniority and authority of G12. He suggested
termination of escape attempts and return to a safer cavity; he led
N12 and Q12 on a final search for water, found a gallon, and rationed it, he suggested drinking urine and tried it first, he supported
and encouraged those in despair, he stayed by the air pipe, heard
the rescuers working, and organized shouting through the pipe until
they were heard; and he demanded and got water, coffee, and soup
piped in. He gave the following account of this·

> I told them that we needed water and they said,
> "You can't have water. Just lay down and take it easy."
> I said, "Look, we got to have water, if we don't get
> water you might as well quit digging." They said,
> "The doctor said you can't have it." And I said, "What
> doctor? You put the doctor on the end of that pipe till
> I talk to him." I told him, "You had better get a small
> pipe through here and give us some water because we
> have a couple of fellows who are hurt and they need
> water and we need water too. We just have to have it."
> I told him what we were doing, and I said, "We can't
> take much of that." So he said, "all right," they
> would put a pipe through...So finally they did get it
> through...Then he told me, "You tell them to take a
> mouthful and count five hundred." So when I got in
> there and told them that, I said, "Here's the water
> go ahead and make a pig of yourself if you want to."
> Apparently nobody did because nobody was sick. Well,
> I knew myself, I didn't do no counting when I got a hold

of it. I got a mouthful and kind of rinsed it around a
bit and then I took a good slug of it. I figured then I
just laid back and let that trickle down...I had taken
the small line in and over top of the pans...and I'd
held the pipe over there so whatever was spilled went
in the pan. I told them, I told the boys that if we got
stuck, if something happened out there and they couldn't
get in, we'd have lots of moisture right there in the pan.
We could dive into that.

The group of six. Three different miners played a leader's role on different occasions in the group of six. Immediately after the bump, other miners looked to C6 because of the danger of gas. C6 was a senior miner, had been trapped in the 1956 explosion, and had been trained as a dreagerman. A6 said,

C6 was a dreagerman and I figured I would stay
close to him--I figured I was safe when I was close to
him. He told us to stay put and we stayed put for a
while.

C6 did not maintain the role that was initially defined for him, however, later on he did restrain F6 from impulsively severing the pinned miner's arm with an axe. He also noticed the pinned miner's death and checked his pulse.

E6 seemed to take the initiative after impact and through the escape period. Immediately after the bump, he dug out the partially buried F6, assessed the danger of gas and the plight of the pinned man, and sent for an "official," presumably acting on the basis of training and experience. He led several escape attempts down the wall, exploring every possible way through. On several of these, he and his friend B6 went together, and it was they who found the isolated man, X6. As late as the sixth day, E6 took his friend and, in spite of darkness and exhaustion, tried again. E6 suggested that they drink their urine and that they try eating coal. However, though his hopes remained quite good through the survival period and he was a willing helper in beating the pans, he was not active in maintaining group morale.

During the survival period, D6 clearly emerged as the leader of the group of six. Immediately following the bump, D6 went to the miner whose arm was crushed and pinned by a pack. He ministered to this man's needs as far as possible and influenced the group not to remove the man's arm. He took little part in the escape activities, however. D6 suggested and supervised the rationing of the water. He decided about the second or third day to beat on the pans and air pipe every hour on the hour in order to announce their presence to the rescuers. He had the help of the more capable members of the group in this operation. He watched the weaker members of the group and when they could not take their turns at "beating the pans" he filled in. In order to dispel despondency, he told jokes and recalled amusing incidents concerning his own life and family. He sang songs and led the group in singing the Old Rugged Cross. He was conscious at all times of the "shaky" man (F6) and encouraged him with positive support: "Always laughing, I would say, 'You will last for another day.'" D6's will to live and his belief in "deliverance by God" seemed to seep into the group during the long survival period. D6 confirmed the death of the pinned miner on the sixth day.

D6 appeared to make full use of his personality assets to promote a type of inspirational group therapy which proved most effective during the extremely stressful and prolonged survival period.

Sociopsychological and Physiological Effects

Hope and Despair

While despair and apathy were common during the later days of entrapment, there were few if any times when at least one miner was not expressing hope of rescue. It may be a characteristic tendency of groups under stress that not all members despair simultaneously.*

The miners' interviews suggest that the expression of despair had a positive and necessary function in the circumstances. If most members of the group expressed despair, one member seemed to be forced to take the role of the optimist:

> I said, "We might as well make up our mind we are here to stay and we are going to go. We will all go together." And he said, "The good Lord didn't keep us here this long to let us go now."

---

*A similar phenomenon has been noted by Perry, Silber, and Bloch (1956). They concluded that a family under stress permits only some of its members to be seen as disturbed, although other members are equally upset.

In another instance, a miner explained to the interviewer "I said, 'They will get us out.' I never thought they would. I was just telling him that."

The ability to shift roles, from voicing despair to hope, is well illustrated in the following conversation:

> I said, "Well, boy, I don't know. It don't look good." I said, "I don't believe we'll make it." I said, "The sound is not getting any handier." And he said, "No." And...he started to cry. And I said, "Boy"--I had tears in my eyes--but I said, "Don't cry, we need all our strength." That's what I said. "It's only three or four days we been here now, five or six days," I said. And I said, "I think I got strength enough yet for a couple more days and maybe more." So he said, "All right...I'll stop." And we talked there quite a while.

Much of the period of waiting was filled with conversation about neutral, nonevocative subjects. "I didn't talk much about my family, although I thought about them." No one talked of fellow workers who had been killed.

The interviewer asked the miners what they thought about while trapped. For the most part, their replies were about fairly trivial recent events, bits of finished and unfinished business ("I got a couple of pieces of hose to put on the heater of the car...It was $1.40, and I didn't pay them, and I didn't leave no note for my wife to see that I owed them this $1.40, and I worried about them not getting their pay."), about hunting and plans for hunting trips. Many recognized these thoughts as cover thoughts for their real preoccupations: their families and the possibility of rescue. These latter thoughts evoked despair and the men attempted to suppress them:

> I was thinking of hunting, and anything like that, you know. Anything at all to take my thoughts away from my family. I sure found that that works. You think of your family and then you start filling up (crying).

> I thought about a little bit of everything. Mostly thought of my family. But I found I was going to fill up a little bit, so I put it out.

> ...I tried to keep too many thoughts from coming
> into my mind, because I knew if I let myself think
> what I could have thought, I probably would have got
> hysterical...If I had let myself go, I suppose I could
> have broken down any time. That is one reason why
> I didn't let myself think too much about getting out.

There was a taboo against crying·

> So he had a little breakdown. I told him, I said,
> "Don't you start that stuff here. You just forget about
> that stuff." Because I figured if it started, it would
> spread around.

Nevertheless, most of the men had a soundless, private cry. "I had a couple of little sniffs...and I think nearly all the other chaps had a couple." "At times, tears used to come into my eyes--well up in my eyes. But I never broke down, I never cried."

Expectations and Behavior

The miners were trapped in an environment with which they were familiar. Special skills and knowledge had been gained through their daily working experience in the mine over many years. It might be expected that this knowledge would permit the miner to anticipate the problems he would face in a mine disaster and to deal with them efficiently, in contrast, for instance, to people suddenly hit by a hurricane.

To some extent this was so. All the miners expected gas following the bump and took the proper precautions against it. "We was afraid of gas and kept our heads down." When nontrapped miners were asked what they would have done had they been trapped, they also talked of precautions against gas:

> The worst thing I could see there would be the gas,
> to try to protect yourself because you can't smell it,
> you can't see it. And you would just have to go until
> you felt yourself getting weaker, because if the gas gets
> you, you gradually get weaker.

Other precautions were not taken. The trapped men gave little thought to the preservation of the vital water supply during their preoccupation with escape. Much more care was given to the preservation of the limited supply of light.

The interviews, moreover, made frequent reference to expectations that were not fulfilled. Some of these expectations were very real and disturbing to the trapped men:

> Oh, it was terrible...The smell (of the dead bodies), that is what I was scared of. The stuff might poison us, you know. The fumes of that. I was scared of that too. Of course, you can't tell yet what it might do to us.

Despite the miners' knowledge of the social code that the rescuers work continuously until the last man is out, they still feared that their rescuers would leave them in the mine:

> Different times we heard a picking noise like someone scratching. And everytime it would stop, we would think, "Now they're done."

> They'll take the weekend off, or give up.

The miners' expert knowledge created difficulties when that knowledge was unreliable under the new critical conditions. Although all trapped miners paid close attention to all sounds, they misread the significance of these because they used as their point of reference the mine as it normally operated. The bump had broken pipes, blocked passages, and disrupted other conductors of sound. Hence, sounds of rescue produced such definitions as

> I said, "Now, boys, they are working away on the level. The condition of this wall means they will be a month getting us out of here." And that is why I gave up hope.

> I am an experienced miner, and I was trying to work it out from the sounds. I had been on rescue work in different places and I knew just what they should be doing. And I was wrong. I was 100 per cent wrong.

The trapped miners had been unsure of their ability to play their role adequately in the critical situation. "I've often said to my wife before I went to work, 'If something happened and I was trapped in the mine, I would go stark raving mad.'" In their interviews, many expressed astonishment and pride that they had managed to remain under control.

This expectation that a trapped man would be unable to carry out his role was widespread. Nontrapped miners, interviewed after the trapped miners had been rescued and their experiences made known, expressed the same anticipation of loss of control had they been trapped. Even with knowledge that the trapped men had remained under control, 60 per cent of the nontrapped miners mentioned fear of losing sanity: "I'd just go crazy, that's all."

Forty per cent of the nontrapped miners did not know how they would have behaved had they been trapped

> It's a hard question. I don't know. To be trapped there, I think it would be impossible for anyone to say what they would do. I couldn't picture what I would do. No, I absolutely couldn't. I never had any experience that way at all.

These expectations of how a man would behave when trapped were in marked contrast to the planned behavior of the wives during the crisis. The wives' expectations and definitions of the emergency situation prepared them for the role they were to play (Chapter 3).

The role of a trapped miner's wife was a public, structured, and active role, played in conjunction with less involved people. The role of a trapped miner was incompletely structured, lacking in support from less involved people, and it contained many unknown qualities. This suggests definite limitations to which some disaster roles can be anticipated, and, as a consequence, a limit on the amount of effective preparation that an individual can make for a crisis, even one for which he has considerable experience and training.

Physical Symptoms

Tables 3 and 4 show the categories of symptoms reported by the group of twelve and the group of six during their entrapment. Descriptive examples of symptoms used under each heading in the tables are

(a) Respiratory symptoms: episodes of hyperventilation; dyspnea on exertion or at rest,

(b) Gastrointestinal symptoms. "butterflies in the stomach"; nausea; gagging, vomiting; diarrhea, cramping pain in the abdomen,

(c) Cardiovascular symptoms: palpitations; consciousness of heart,

(d) Genitourinary symptoms: frequent micturition; "scalding"; dysuria,

(e) Skin symptoms: "goose pimples," paraesthesia or "numbness," "pins and needles," "skin felt like that of a corpse,

(f) Musculoskeletal symptoms: cramping in the muscles of the legs and arms or trunk,

(g) Headaches: mostly described as frontal headaches.

TABLE 3

Psychophysiological Symptoms During Entrapment:
Group of Twelve

Miner	Respiratory system	Gastrointestinal system	Cardiovascular system	Genitourinary system	Skin	Musculoskeletal system	Headache
G12	P	-	P	-	-	-	-
H12	-	-	-	-	P	-	-
I12	-	-	P	-	P	P	P
J12	P	P	-	-	-	-	-
K12	P	P	-	P	P	P	-
L12	-	-	P	-	P	-	P
M12	-	-	P	-	P	P	P
N12	P	-	P	-	P	-	P
O12	-	-	-	-	-	-	-
P12	-	P	-	-	-	P	-
Q12	P	P	-	-	-	-	P
R12	P	-	P	P	P	P	P

Note: P = presence of symptom reported; - = presence of symptom not reported.

Each trapped miner reported a number of symptoms. It is difficult to assess the origin of the symptoms. All of the men were subjected to starvation, dehydration, disturbance of electrolytic balance, sensory deprivation, noxious gases, and, of course, severe stress. Several were seriously injured in the bump; all were bruised. Pre-existing illnesses and infection must also be taken into account. No doubt many symptoms were signs of anxiety.

TABLE 4

Psychophysiological Symptoms During Entrapment: Group of Six

Miner	Respiratory system	Gastrointestinal system	Cardiovascular system	Genitourinary system	Skin	Musculoskeletal system	Headache
A6	-	P	-	-	P	-	-
B6	-	-	P	-	-	-	-
C6	-	-	-	-	P	P	-
D6	-	-	-	-	P	P	-
E6	-	P	P	-	P	-	P
F6	-	-	-	-	-	-	-

Note: P = presence of symptom reported; - = presence of symptom not reported.

It is noteworthy that there were fewer symptoms reported by the group of six than by the group of twelve. The group of six were trapped 2 1/2 days longer and they were exposed for 5 1/2 days to a dying man who was sometimes delirious and whom they were powerless to help. This man's cries for water and his pleas for them to amputate his arm were very stressful to the group of six. This unusual stress may have diverted their attention away from their own complaints. On the other hand, the stronger social controls in this small group (see Chapter 5) may have exerted a repressive effect on awareness of complaints.

Ten of the 18 trapped men did not have a bowel movement during the entire period.

The men seem to have slept for short, irregular periods of an hour or two, with equally long periods of full or half consciousness:

> I would lay back and I would pray and I would fall asleep. I would sleep for an hour. Then you would wake up and start worrying what was going on again.

The first few days I couldn't sleep at all. The only time I did sleep was after we had got the water and soup into us.

Nobody slept all at once. I don't think they did, because any time I woke up I would always say, "Is there anybody awake?" Generally one or two of the boys would say, "Yeah, we are."

The results of the physical examination that the trapped miners received on rescue are reported in Appendix B.

Deprivation of Sensory Stimulation

Both groups of trapped men spent the greater period of their entrapment in total darkness: the group of twelve for 3 1/2 days and the group of six for 5 1/2 days. Experimental visual deprivation has produced hallucinations (Heron, Doane, & Scott, 1956, Solomon, Leiderman, Mendelson, & Wexler, 1957), and almost all the trapped miners reported hallucinations, though they were described in different terms. The following six are typical descriptions:

(a) Like neon lights and everything. More or less the nerves of our eyes. You could imagine you could see dust and couldn't see nothing. It occurred pretty well towards the last of it.

(b) I have seen more lights and there were times that my eyes were shining like headlights in a car. Just come right up like that. The first time I experienced it--seen these lights, yellow and blue--I called them headlights and I would say, "Boys, my headlights are shining." I would almost take my oath I could see it, see them packs and the stone wall. I would open my eyes for a while and they would go away. But after a little while they would come again. I just thought, well, that's because it's so dark, that's what is making them go.

(c) I began to see a red--a yellow glow. I don't know what it was but perhaps it was the fact that we wanted to see light. We were so long in the dark, we persuaded ourselves we could see a yellow light.

(d) Well, it was just like lights flickering in your eyes. Your eyes would open and still it was flickering in your eyes. I was laying with my eyes closed and I would swear somebody had a light on where we were. It just seemed like everything was lit up. Imagination I figured it was. But we had a lot of troubles at the last. We had a lot of it like that. Everybody was saying they had that trouble.

(e) When we lay down, we would see like little spots, like little fellows running away. And different times, it looked as if there were neon signs flashing on and off in the dark. As soon as you would close your eyelids you would see these little lights and things dancing in front of your eyes.

(f) I could see lights in front of my eyes. I'll tell you what they reminded me of--these kaleidoscopes, I think they call them, with all the little cut glass in. It was like one of these going all the time.

One man reported a body illusion. He had a fracture of the lower third of the right femur, which caused his leg to be painful and swollen. He said,

I began to think of my leg as if it wasn't part of me, and I would keep on saying, "My poor old leg! My poor old leg!" I began to feel sorry for it. But I didn't think it was part of me at the time. Maybe that was a good thing.

Apart from these phenomena, no other effects of deprivation of sensory stimulation were reported. It is noteworthy that starvation, dehydration, fatigue, an abnormal percentage of methane and carbon dioxide, apprehension, and anxiety, along with deprivation of visual stimulation for up to five days did not produce more florid and more organized hallucinations. The fact that the men were not isolated from one another may have been an important counterfactor. It is noteworthy, too, that the men's discussion of their hallucinations while they were experiencing them did not seem to produce a common form of hallucination.

Recent experimental studies indicate that cognitive functions are impaired after as little as 24 hours of sensory deprivation, with some evidence that the effects last up to two days (Doane, Mahatoo, Heron, & Scott, 1959; Scott, Bexton, Heron, & Doane, 1959). Since miners were not tested until 6 to 23 days after their rescue, it is unlikely that any effects of perceptual isolation were reflected in their cognitive test performance.

CHAPTER 5

BEHAVIOR OF THE TRAPPED MINERS
A QUANTITATIVE ANALYSIS

The behavior and the experiences of the trapped miners during their entrapment have been discussed in general terms in the preceding chapter. Now the same phenomena are considered from a different point of view. This chapter systematically describes one quantifiable aspect of the miners' behavior: their initiations.

An initiation, as defined in this study, is an act that originated an extended sequence of behavior.

By counting initiations through content analysis of the interview material, it is possible to consider some of the actions of the miners in a specific and systematic way so that this aspect of their behavior can be compared. With such a measure, it is possible to find out how equally all members of a group initiated action and if the frequency of a miner's initiating was constant throughout the entire entrapment or varied during different periods of this experience.

One expects great differences in the accounts of such a prolonged and harrowing experience. The spontaneous story obviously reflected selective recall, conscious and unconscious reconstruction of the events, and deficiency in original perception. Although there was no way of accurately and objectively judging the relative validity of accounts, there was no reason to discount the validity of any account. Each point of view and description of events was regarded as adding some part to the picture of the total experience. Therefore, the analysis considered the number of initiations each miner attributed to himself, the number he attributed to others, and the number others attributed to him.

The individual was not asked to rate the persons in the group, rather, his spontaneous story was analyzed to yield a score for himself and his companions. Each speaker in telling his story credited himself with the initiation of certain lines of action or with making suggestions upon which the group acted: "I dug him out and he thanked me." "I seen we had to have help so I said to B6, 'Go out the timber road and get help'...He said, 'All right.'" He also credited others by name for making suggestions or initiating actions "D6 said we should ration the water out...and he did." Other initiations were credited to specific but unnamed persons "Somebody said--I don't remember his name--but somebody said, 'Let's pound again.' So two of us..." Still others were credited to the total group or subgroup as "we" and "they" "We got this saw and we sawed our way out to the bull wheel," and, "When they got the hole through and the air was coming through, they hollered for us to come down." A point was scored each time the subject mentioned an initiation, and the point was attributed to either the speaker, the named initator, or one of the three general categories, "we," "they," or "unspecified."

Thus, the spontaneous story of each member of the group was translated into a point-score indicating the speaker's view of the initiation pattern. The following kinds of initiation scores were used in the analysis:

(a) <u>Initiation Perception</u> the total number of times the miner mentioned initiations in his interview, including the initiations the miner attributed to himself, to other members of the group, and to the whole group or subgroups.

(b) <u>"I" Initiation</u> the number of times a miner mentioned initiations that he attributed to himself.

(c) <u>Group Evaluation</u> the total number of times all other members of the group mentioned initiations attributed to one particular individual, exclusive of the initiations with which the individual credited himself. (If two miners mentioned the same initiation, the score is two.)

Although numerical scores are included in the tables for this chapter, the analysis is based on rank orders. The rank orders are highly reliable and less affected by the account of a single respondent than are numerical scores. All of eight judges, independently scoring the interviews, ranked the miners in the same order. Even in the case of untaped interviews, though they may be incomplete in detail, it should be borne in mind that the interviewer when dictating the untaped interview recounted the highlights and major activities reported in the respondent's story. Thus, the proportion of initiations that the respondent attributed to other miners would tend to be preserved in the interviewer's dictated account. Consequently, untaped interviews are unlikely to affect rank orders of Group Evaluation. It should also be noted that the number of initiations mentioned in each interview did not show any direct relationship with the length of the interview. Many long interviews contained fewer initiations than short interviews. Interview probes rarely increased the number of initiation perceptions.

As an additional check on the reliability of the rank order from the Group Evaluation score, two methods were developed. The first method analyzed the initiations mentioned in each interview to find the highest initiator as recognized by that respondent. A point was credited to the highest initiator derived from each interview. The group members were then ranked according to the number of credits each received. This rank order was identical with that from the Group Evaluation score.

The second method analyzed the initiations in each interview to find the lowest initiator as recognized by that respondent. A point was credited to the lowest initiator as derived from each interview. The group members were ranked according to the number of credits each received. This rank order was the inverse of the high initiator rank order and the rank order from the Group Evaluation score.

Initiations

The Group of Six

Table 5 lists the miners of the group of six in rank order of Initiation Perception, "I" Initiation, and Group Evaluation scores.

There was a wide range in the number of initiations identified by each man--from a high score of 54 to a low score of 11. The story of A6, who is ranked lowest on initiation perception, was not recorded on tape due to difficulties with the machine. Instead, it was recorded immediately afterwards by the interviewer from notes taken during the interview. The interviewer was experienced, and long passages of direct quotations were included in the interview material with little summarizing or condensation. However, it must be noted that this particular interview is atypical.

Each miner, recounting the story of the entrapment, attributed more initiations to himself than to anyone else. As the story was given from the speaker's point of view, this is hardly surprising. There were considerable individual differences in these "I" Initiation

TABLE 5

Rank Order by Initiation Perception,
"I" Initiation, and Group Evaluation Scores:
Group of Six

Initiation perception		"I" initiation		Group evaluation	
Miner	Score	Miner	Score	Miner	Score
E6	54	E6	28	D6	13
C6	29	D6	16	E6	12
F6	26	F6	12	A6	10
D6	19	A6	8[a]	B6	8
B6	12	C6	7	C6	2
A6	11[a]	B6	6	F6	2

[a]Based on untaped interview.

scores: one member recounted 28 actions and decisions he had initiated, while another only attributed 6 to himself. The range, however, was not as great as that of the Initiation Perception.

Scores on Initiation Perception and "I" Initiation were derived from individual interviews. The Group Evaluation scores, on the other hand, were derived for each person in the group from the interviews of all the members of the group and were, therefore, the most important in the assessment of group structure. As can be seen in Table 5, 2 of the 6 miners were credited with a mere two initiations in the accounts of their companions, considerably fewer than the initiations credited to the other 4.

The Group of Twelve

The scores from the initiation analysis of the interviews of the miners in the group of twelve are listed in rank order in Table 6.

Four of the 12 interviews were not directly tape recorded. Machine difficulties required the interviewer to make as complete notes as possible during the interviews of miners G12, O12, P12, and R12. The lack of detail of interviews recorded in this fashion undoubtedly depressed the scores of G12, O12, P12, and R12 on Initiation Perception and "I" Initiation.

TABLE 6

Rank Order by Initiation Perception,
"I" Initiation, and Group Evaluation Scores:
Group of Twelve

Initiation perception		"I" initiation		Group evaluation	
Miner	Score	Miner	Score	Miner	Score
Q12	77	K12	30	G12	15
K12	56	Q12	19	K12	11
N12	30	N12	10	M12	11
M12	27	M12	7	I12	9
H12	24	O12	6[a]	Q12	8
I12	19	G12	5[a]	L12	7
J12	16	H12	5	N12	6
L12	15	I12	4	P12	4
O12	14[a]	J12	3	H12	2
G12	9[a]	L12	2	J12	0
R12	7[a]	P12	2[a]	O12	0
P12	4[a]	R12	1[a]	R12	0

[a]Based on untaped interview.

Again, as in the group of six, there were wide ranges in scores on Initiation Perception, "I" Initiation, and Group Evaluation. Some miners attributed few initiations either to themselves or to their companions. Those who credited themselves with many initiations also credited many to others. Miners J12, O12, and R12 were given no credit by their companions for initiation; one man, G12, was given a prominently high score.

Comparison of the Two Groups

The group of six and the group of twelve differed in a number of ways, e.g., time underground, spatial proximity, and size of group. Both groups lacked formal structure and leadership when the collapse of the mine isolated them. As soon as the first initiations were made, an informal structure began to emerge. This informal structure showed different characteristics for the two groups.

Three of the miners in the group of twelve were credited with no initiations by their companions (Table 6); no member of the group of six was credited with fewer than two initiations (Table 5). Thus, in the larger group there were more men who played no significant role in the group's activities.

That the initiation pattern changes with the size of the group is well documented in the experimental literature (Kelly, 1954). For example, Bales (Bales, Stradtbeck, Mills, & Roseborough, 1951), using a "basic initiation rank," found that the proportion of very infrequent contributors to group interaction increases as the size of the group increases. Both Carter (Carter, Haythorn, Shriver, & Lanzetta, 1951) and Gibb (1951) suggest that in larger groups the amount of freedom is not sufficient to accommodate all the group members.

The group of twelve more often attributed initiations to "we" or to "they," or left the initiator unspecified. These categories accounted for 131 (43.9 per cent) of the 298 initiations reported by the group of twelve and 27 (17.8 per cent) of the 151 initiations reported by the group of six. The difference between these percentages is highly significant statistically.

Again, this is probably a function of group size. With an increase in the size of the group there is greater opportunity for forming subgroup coalitions, and more of the activities of the group become depersonalized.

Initiations in Time Sequence

The initiation pattern of a group tends to change with time. Bales (1950) and Gibb (1954) have noted general tendencies toward a shift in emphasis of task and goal within ongoing groups. Riecken and Homans (1954) have discussed phases that groups go through in solving a problem and noted that as a group becomes more accustomed to working together less interaction is spent on the task at hand and more is spent on affective reactions.

The change of initiation patterns through time is particularly pertinent for the trapped miners because their period of entrapment seems to break down into phases, each one having its own rather distinct problems.

The Group of Six

Two periods could be distinguished in the accounts of the six miners.

Escape period. Most of the time during the first three days of entrapment was spent in trying to escape. During this period there was a great deal of physical activity and exploring as possibilities of escape were probed.

Survival period. After about three days of fruitless activity, the miners were without light and their water was nearly depleted. They now realized that they were unable to free themselves by their own efforts, therefore, most of the remaining five days before rescue were devoted to survival, and a different set of decisions and actions was brought into play.

Pinned miner episode. Part of the first five days was spent dealing with a crisis quite distinct from escape or survival. It concerned a seventh miner caught by his arm in one of the wooden packs that held up the roof of the passageway. He was in pain, and the arm could not be released. The group had to decide what to do. The alternatives suggested were to remove the arm, risking an amateur operation with an axe and the possibility of the man's bleeding to death or becoming uncontrollably violent and threatening the existence of the whole group, or to leave the man pinned as he was, against the wishes of the man himself. Eventually, their decision was to leave him as he was, and the miner died on the fifth day of entrapment.

Although the escape period lasted for only the first three of the 8 1/2 days, a majority of the initiations were reported for the escape period. Eighty-four initiations occurred in the accounts of the escape period, 48 in the accounts of the ensuing survival period, and 19 in the accounts of the pinned miner episode.

The few initiations reported in the pinned miner episode had no relation to the amount of time spent discussing the pinned miner in the interviews. All discussed him at length, not in terms of action or initiation, but rather weighing the pros and cons of the decision not to amputate his arm. Almost half of the initiations that were referred to were attributed to "we" (a much larger proportion than in either the escape or survival periods). Only one miner attributed an initiation to another miner in his account of this episode.

Table 7 lists the group evaluation of each miner's initiations during the escape period and the survival period. It is clear that in the reports of their companions none of the men were consistently high initiators during the entire period of entrapment. Instead, three (E6, A6, and B6) were reported active during the escape period, and another (D6), who received only one credit for an initiation during the escape period, received almost all the credits for initiation during the survival period. Initiations were spread more evenly among the six miners for the escape period than for the survival period. The escape and survival periods, then, were distinct, not only in the kinds and amount of activities carried on, but also in the individuals who initiated these activities.

TABLE 7

Rank Order by Group Evaluation Score
in the Escape Period, the Pinned Miner Episode,
and the Survival Period:
Group of Six

Escape period		Pinned miner episode		Survival period	
Miner	Score	Miner	Score	Miner	Score
E6	11	D6	1	D6	11
A6	9			B6	2
B6	6			A6	1
C6	2			E6	1
F6	2			C6	0
D6	1			F6	0

The same shift was less clearly reflected in the "I" Initiation score for the escape and survival periods (Table 8). E6 attributed almost as many initiations to himself during the escape period as the other five miners combined. For the survival period, D6, the high man in Group Evaluation during this period, barely displaced E6 as the highest in "I" Initiation. On the whole, "I" Initiation, or self-evaluation, was not closely related to Group Evaluation.

The total number of initiations each man reported remained quite constant for the escape and survival periods (Table 9) and largely independent of both Group Evaluation and "I" Initiation.

The Group of Twelve

There was no major situation during the entrapment of the group of twelve, similar to the pinned miner episode in the group of six, that could be distinguished from initial attempts to escape or later attempts to survive until rescued. Therefore, the initiation scores of the group of twelve were analyzed separately for the escape and the survival periods.

TABLE 8

Rank Order by "I" Initiation Score
in the Escape Period, the Pinned Miner Episode,
and the Survival Period:
Group of Six

Escape period		Pinned miner episode		Survival period	
Miner	Score	Miner	Score	Miner	Score
E6	19	A6	2[a]	D6	9
B6	5	C6	2	E6	8
D6	5	D6	2	F6	7
A6	4[a]	F6	2	A6	2[a]
C6	4	E6	1	B6	1
F6	3	B6	0	C6	1

[a]Based on untaped interview.

TABLE 9

Rank Order by Initiation Perception Score
in the Escape Period, the Pinned Miner Episode,
and the Survival Period:
Group of Six

Escape period		Pinned miner episode		Survival period	
Miner	Score	Miner	Score	Miner	Score
E6	34	E6	5	E6	15
C6	18	F6	5	F6	12
B6	9	C6	4	D6	9
F6	9	D6	3	C6	7
A6	7[a]	A6	2[a]	B6	3
D6	7	B6	0	A6	2[a]

[a]Based on untaped interview.

   As in the group of six, the escape period lasted for about three days. Rescue came after nearly 6 1/2 days of entrapment. Also as in the group of six, most initiations were reported for the escape period. The accounts for the first three days yielded 212 initiations; accounts for the last three, 86 initiations. The activity of the group, then, seems to have changed markedly after it was realized that attempts to escape were futile.

The Group Evaluation of each man's initiations during the escape and survival periods is shown in Table 10. The Group Evaluation scores of the larger group of miners were similar to those of the smaller group in that the patterns of evaluations for escape and survival periods differ. G12 stood out during the escape period with 13 initiations attributed to him, 4 more than the next highest man; he received credit for only 2 initiations during the survival period. K12 was credited with only one initiation during the escape period; with 10 initiations during the survival period, he had twice as many as anyone else in the group. As in the group of six, more men shared in the initiations for the escape period than for the survival period.

TABLE 10

Rank Order by Group Evaluation Score
in the Escape Period and the Survival Period:
Group of Twelve

Escape period		Survival period	
Miner	Score	Miner	Score
G12	13	K12	10
I12	9	N12	5
M12	9	Q12	5
L12	7	G12	2
P12	3	M12	2
Q12	3	P12	1
H12	2	H12	0
K12	1	I12	0
N12	1	J12	0
J12	0	L12	0
O12	0	O12	0
R12	0	R12	0

Tables 11 and 12 give the "I" Initiation and Initiation Perception scores for the two periods. Unfortunately, the interview with G12, the high man in Group Evaluation for the escape period, was not directly tape recorded. This undoubtedly depressed his "I" Initiation and Initiation Perception scores. For the survival period, K12 led as markedly in "I" Initiation as he did in Group Evaluation.

TABLE 11

Rank Order by "I" Initiation Score
in the Escape Period and the Survival Period:
Group of Twelve

Escape period		Survival period	
Miner	Score	Miner	Score
K12	17	K12	13
Q12	15	N12	4
M12	7	Q12	4
N12	6	H12	3
G12	4[a]	O12	3[a]
I12	3	G12	1[a]
J12	3	I12	1
O12	3[a]	L12	1
H12	2	J12	0
P12	2[a]	M12	0
L12	1	P12	0[a]
R12	1[a]	R12	0[a]

[a]Based on untaped interview.

TABLE 12

Rank Order by Initiation Perception Score
in the Escape Period and the Survival Period:
Group of Twelve

Escape period		Survival period	
Miner	Score	Miner	Score
Q12	64	K12	18
K12	38	H12	15
M12	24	Q12	13
N12	21	N12	9
I12	13	O12	7[a]
J12	12	I12	6
H12	9	L12	6
L12	9	J12	4
O12	7[a]	G12	3[a]
G12	6[a]	M12	3
R12	5[a]	R12	2[a]
P12	4[a]	P12	0[a]

[a]Based on untaped interview.

As in the group of six, the orders of Initiation Perception and "I" Initiation scores varied less from the escape to survival periods than did Group Evaluation. The miners in both groups tended to report consistently in terms of action or nonaction, and to report themselves as consistent initiators, either high or low, but they distributed their credits for initiations to their companions differently for the escape and the survival periods.

Discussion

Leadership and Initiation

The defining characteristics of leadership have been variously identified, but presumably one basic and unique function of the leader is initiation: the origination and facilitation of, or resistance to, new ideas and practices (Bales, 1953, Halpin & Winer, 1952; Horsfall & Arensberg, 1949). Halpin and Winer (1952) found a close relationship between leadership ratings and amount of initiating. The high initiators in the two groups of trapped miners, therefore, can be identified conventionally as leaders.

A number of different techniques for the rating of leadership are found in the experimental literature: rating by extra-group observers, by the participants in the group activity, and by self-appraisal of group members. Generally the results of these different methods have been found not to be highly related to each other. In this study, there is information on initiation from both self-appraisal ("I" Initiation) and Group Evaluation. The relationship between the two is slight. It has also been noted that the "I" Initiation score tended to remain constant although the Group Evaluation changed considerably from one time period to another. The two scores, then, tended to measure different things.

An "I" Initiation score reflects personality characteristics that may not be related to leadership, particularly the individual's habitual style of narration. The narrative style of some interviews displayed a series of finely distinguished decisions, orders, and actions, yielding a high "I" Initiation score. For example, the following is a characteristic sequence in E6's narration·

> I started down the wall, and I heard someone hollering "Help!"; so I went down and there was Joe in the pack and just below I heard somebody hollering "Help!" so I went down below and here was Jim. He was buried about round the waist. I dug him out, and he thanked me, and I said, "Don't mention that, boy." And we

> came back up--we came up, and I looked at Joe's arm
> and I seen we had to have help so I yelled to Archie,
> "Go out the timber road and get help"; and I said,
> "Bring an official in with you, its going to be gassy in
> here."

Other miners recounted their experiences in terms of thoughts rather than action, thus earning low "I" Initiation scores. For example, D6, who had a high group evaluation score, tended to speak in a frame of reference of "I thought" or "we figured".

> Well, I thought about getting up, I thought the level was
> our best bet...Well, the way I figured, we didn't have
> the tools then, the way I figured, if we could have re-
> leased...

Such idiosyncracies in the individual interviews may have little to do with leadership characteristics, but greatly influence an individual's "I" Initiation score. For this reason alone the Group Evaluation is to be preferred as a measure of leadership.

Group Evaluation is also preferable on theoretical grounds. As Gibb (1954) points out, man emerges as a leader not simply because of particular personality traits but because of his ability to satisfy group needs. The Group Evaluation score seems to be a measure of the individual's standing, which depends not on the individual's initiations as such but on the extent to which his companions perceive him as having made these initiations. Gibb (1954) adds that "his standing in turn is dependent not upon possession of these special qualities as such, but upon the extent to which his fellows perceive him as having these qualities."

This dimension of perception of leadership is largely absent from the "I" Initiation scores. The "I" Initiation score is a measure of the number of initiations the individual claims to have made. Each miner may well have initiated as many times as he said he did--and probably at a fairly constant level throughout entrapment. However, only some of these initiations were seen, recognized, and remembered by other members of the group, who would perceive only those initiations required by them at a particular time to achieve a particular goal in a particular situation. If this is so, the Group Evaluation score reflects a social evaluation relative to other members and relative to the situation, and therefore this evaluation approaches more closely the conventional definition of leadership.

The Shift in Leadership

The leadership of a small, traditionless group may shift frequently as the group moves from one phase of activity to another (Gibb, 1947). At any one time, a group member achieves the status of a group leader in proportion to his participation in the group activity and his demonstration of capacity to contribute more than others to the achievement of the common goal. As the activity changes, the roles required of a leader change, and, because of individual differences among group members, the likelihood is that different members will be perceived as filling these roles best. Carter (1953) suggests that

> there are probably families of situations for which leadership is fairly general for any task falling in that family, but there will be other families in which leadership requirements will be fairly independent of those in the first family of situations.

Two such independent families of situations have been identified by Bales (1953): task-oriented (instrumental-adaptive) and emotion-oriented (expressive-integrative), each with its peculiar leadership requirements.

Bale's classification of situations fits the entrapment of the miners. The miners' interviews all indicate a change in the major goal of both groups about three days after the bump. The first phase was an attempt to escape, it was characterized by vigorous action and little concern with keeping up morale. While most miners made initiations during this period, one man in each group, E6 in the group of six and G12 in the group of twelve, was prominently high in initiations and can therefore be identified as the leader. These two men were task leaders although the task was widely shared among their companions.

In the survival period, the miners concentrated on waiting and surviving, with a little physical action maintained in pounding on the pipes to attract the attention of rescuers. In both groups during this period a new leader emerged. K12 in the group of twelve was credited with almost as many initiations during the survival period as the rest of his group combined, while D6 in his group was credited with almost three times as many as the rest of his group combined. They were the emotion-oriented leaders of their groups and their role was not widely shared.

Presumably, each of the leaders--the task-oriented leaders and the emotion-oriented leaders--emerged in their role because of appropriate qualities. Certainly a number of the trapped men were precluded by injury from leadership, at least during the escape period and perhaps during the survival period as well. C6 and F6 in the group of six and O12, P12, and R12 in the group of twelve were all injured in some degree during the bump and were all low initiators throughout.

Apart from this, the few actuarial characteristics that were investigated proved to be unrelated to the number of initiations: neither age nor years of experience in mining and type of work performed in the mine differentiated the task-oriented leaders or emotion-oriented leaders from their companions. The problem of leadership characteristics is considered more fully in Chapter 8.

CHAPTER 6

PSYCHOLOGICAL DATA ON TRAPPED AND NONTRAPPED MINERS

The 19 trapped miners were all given a brief battery of psychological tests within 5 to 23 days after being rescued. The "control" group of 12 off-shift miners also was given the same battery from 10 to 17 days after rescue operations were completed. The battery included subtests from the Wechsler Adult Intelligence Scale (WAIS) (Wechsler, 1955), two counting tests (counting backwards from 20 to 1 and counting backwards from 100 by 7's), the Bender-Gestalt Drawing Test (Bender, 1938), Rorschach cards I, III, VI, VIII, and X (Rorschach, 1942) and a Sentence Completion test developed for this study (see Appendix C). Three WAIS subtests were selected as measures of different aspects of intellectual functioning Vocabulary for an estimate of general intelligence (it was not given to 2 of the trapped and 9 of the nontrapped miners because of time limitations), Block Design for its performance nature and as a measure of the ability to abstract, analyze, and integrate the components of a problem, and Digit Span to assess immediate memory, attention, and concentration. The two counting tests were included as measures of concentration without involvement of immediate memory. The Bender-Gestalt was utilized as a measure of visual-motor coordination, which is sensitive to degree of emotional disturbance. The Rorschach ink blot test was included in the battery as a means of appraising subjects' modes of handling unstructured stimulus situations. The complete test of ten cards was given to 9 subjects, time limitations and subjects' tendency to fatigue led to the decision to use only five cards on the rest of the subjects. A 31-item Sentence Completion test was made up to elicit information about subjects' feelings and attitudes with reference to events and figures involved in the disaster.

In order to carry out a comparative analysis, all of the psychological data were treated quantitatively. Scores on tests of intellectual functions were derived in the standard manner. To quantify the Bender-Gestalt, a simple method of scoring deviations from the original stimulus drawings was devised, with possible scores ranging from 0 to 12. A subject's score was referred to as his Bender-Gestalt (B-G) error score. The scoring method showed high agreement with clinical judgment in the placing of the 31 drawings into good or poor groups, the two methods were in disagreement on the placement of only 2 out of the 31 drawings. Scoring was reliable between independent psychologists, with only one disagreement in the resulting rank order.

The Rorschach protocols were independently scored by a clinical psychologist. Klopfer's method (Klopfer, Ainsworth, Klopfer, & Holt, 1954) was used with the following modifications

(a) Popular (P) was given for those "common" responses used by Rapaport (Rapaport, Gill, & Schafer, 1946). Card I bat, butterfly, bird, bug, Card III people, bird, butterfly, bow tie, hair bow, Card VI animal skin, Card VIII animal, Card X octopuses, crabs, rabbit head, wishbone, two bugs, worms, and caterpillar.

(b) Only main responses and main determinants were scored and used.

(c) Every main response was given a form-level rating: good form (F+) for reasonably accurate percepts of at least popular form level, poor form (F-) for inaccurate or vague responses.

(d) Those responses that involved determinants other than F were assigned determinant scores in the usual manner.

(e) All whole and cut-off whole responses (W) were scored as either good W's (W+) for well-developed and reasonably accurate wholes or poor W's (W-) for inaccurate, disorganized, or vague wholes.

Each subject's Rorschach scores were combined into four categories defined on the basis of clinical experience and perusal of the literature. The first category was labeled "Rorschach control" and was derived by adding the subject's score for F+, W+, P, FC, number of content categories, and one point if the man's record had one mF or Fm. The second category was labeled "Rorschach disorganization" and was made up of a subject's scores for F-, W-, CF, and C, and a point for each m response over and above the one counted under Rorschach control. The third category comprised the sum of a subject's c and H responses. It was labeled the "empathy" score. The empathy scores varied little, so they were not used in group comparisons. The fourth category was the number of responses each miner gave to the five Rorschach cards.

The use of only five of the ten ink blots is not in keeping with standard Rorschach practice. This expedient, however, would seem justified in view of the high split-half reliabilities found by Hertz (1934) and the conclusions of Baugham (1954), Dörken (1950, 1956), and Frank (1939) that a variety of ink blots evoke relatively consistent responses. Moreover, use of half the stimulus cards is legitimate within the appropriate experimental design. The present study was a comparative one with an experimental and a control group, and at a later date both groups were retested. Finally, only a few relatively simple determinant scores were used in the group comparisons.

Nevertheless, it is worth noting that the five cards produced responses reasonably representative of the standard ten cards in those nine miners who were given the complete series of blots. Rorschach control scores for the two halves of the Rorschach were significantly related (r = .87), while the correlation for Rorschach disorganization was just short of significance at the 5 per cent level (r = .64). The part-whole reliability coefficients were .97 and .94 for the two Rorschach measures respectively. These findings indicate that the five-card Rorschach produced results much like a ten-card Rorschach, at least in terms of the Rorschach control and disorganization scores used in this study.

The Sentence Completion responses were quantified by setting up mutually exclusive and exhaustive categories for the responses to each item. The selection of categories was guided by the nature of the subjects' responses. Where feasible, like categories were used for more than one item. For example, for the three items, "My ambition...," "I would like to...," and "What I need...," two main categories were used -- self-reference and economic reference. Each category was defined with a few examples, and two psychologists independently coded the responses of 20 subjects chosen randomly from the total sample of 31. There was an average of just over three categories per item. The judges agreed on the categorization of 84 per cent of the 620 responses.

A prorated IQ was calculated for each subject from his Digit Span and Block Design scale scores. Since performance on both of these subtests might be expected to be impaired under stress, it could be argued that the Vocabulary subtest would have provided a more valid estimate of intelligence. However, the latter was not given to all subjects. Moreover, for the 20 subjects who did do the Vocabulary subtest, the mean scaled score was 7.6, while the mean scaled score for Block Design and Digit Span combined was 6.7. The difference between these two means was not significant $\underline{t}$ = 1.588)

Thus, the reported IQ is presumably not appreciably different from one that might have been based on the Vocabulary subtest. The fact that the miners' average prorated IQ was considerably below the norm of 100 used by Wechsler (1955) is in keeping with the achievement nature and cultural bias of measures of intelligence (Anastasi, 1958; Burnett, Beach, & Sullivan, 1958). Moreover, the education level of the subjects of this study was quite low (mean of grade 6.7), and education tends to correlate with intelligence in the general population (Anastasi, 1958). Also, the stress that all, or almost all, experienced shortly before the testing may have depressed intelligence scores.

Wechsler's scaled Digit Span score is based on a raw score computed by adding the number of digits that a subject remembers forward to the number he remembers backward. In order to give greater weight to the difference between the number remembered forward and the number remembered backward, a difference which may be increased by disturbances like depression or brain damage, a special score was devised. Hereafter referred to as the Digit Efficiency score, it was derived by adding a subject's digits forward to his digits backward, and subtracting from this sum the difference between his digits forward and digits backward. For example, if a subject's digits forward and digits backward were 7 and 4 respectively, his Digit Efficiency score would be: $(7 + 4)-(7 - 4) = 8$.

The final measure used in this analysis was a "psychiatric ego strength" score based on the interviews. Eight attributes of behavior were rated independently by two psychiatrists on ten-point scales, and the sum of these ratings for a given miner constituted his psychiatric ego strength score. The rated attributes were general ability, stability, initiative, persistence, flexibility, integration, approach to problems (from open, practical, direct, and decisive to panicky or withdrawn), and freedom from symptoms and adjustment problems. The psychiatrists' ratings were in fairly good agreement, with a rank correlation of .67 for the 31 trapped and nontrapped miners (significant at the 1 per cent level). The means of the psychiatrists' ratings were used as subjects' ego strength scores. It should be noted that these scores were based on the subjects' own accounts of their history and their behavior in the disaster, no attempt being made to rate these two parts of the interview separately.

In the analysis of the psychological data the following groups of subjects were compared: (a) trapped miners vs. nontrapped miners, (b) younger trapped miners vs. older trapped miners, (c) younger nontrapped miners vs. older nontrapped miners, and

(d) the group of six (trapped 8 1/2 days) vs. the group of twelve (trapped 6 1/2 days). In addition, rank order correlations, corrected for ties (Horn, 1942), were computed between all tests. All differences and relationships significant at a probability level of 5 per cent or less are discussed, as well as a few pattern trends. The coding of Sentence Completion responses into a number of categories resulted in frequencies that were usually too small for tests of significance. The results were, therefore, treated as percentages of members of a group that gave a particular kind of response. A difference between groups of at least 20 per cent was arbitrarily selected as a minimum for reporting and discussing group trends.

While the use of statistical tools like the t-test, Spearman's rank order correlation coefficient, and a probability level of 5 per cent implies that the relevant assumptions for these statistics have been met, this was not always the case in this study. The groups were small: 12 and 19. Though the control group was matched with the experimental group for age, education, income, religion, and marital status, it is clear this list does not include all the variables one would wish to control. Moreover, the experimental group of 19 trapped miners was not selected randomly, and the parameters of the population were unknown. Further, this was a field study under emergency conditions, and the primary group was given by the accident of the bump.

Where the assumptions for a particular statistic were in doubt, the finding was checked with another tool all significant t-test findings were confirmed with the Mann-Whitney U test (Siegel, 1956), which does not require normality of distributions and interval scales. Treated with appropriate caution, a statistical analysis can be expected to sort out the data more reliably and with less bias than an impressionistic and intuitive analysis. The statistical findings are presented to permit the reader to examine critically the sources of interpretative statements and hypotheses. The objective was not to establish conclusions or generalizations, and the statistical data do not yield them. The data yield hypotheses which must be treated only as hypotheses for research rather than as conclusions for practical social engineering application.

Comparison of Trapped and Nontrapped Miners

The two groups of trapped and nontrapped miners were well matched for age and education (Table 13). The groups differed significantly on 2 of the 11 test measures: on Block Design and Bender-Gestalt errors (5 per cent level with t's corrected for inequality of variances), with the trapped group doing better on both tests.

Moreover, the trapped group did somewhat better than the non-trapped group on 9 of the 11 measures, the two exceptions being counting backwards from 20 to 1, and substracting from 100 by 7's. The trapped group also exhibited more variance on the Block Design subtest (F ratio = 4.586, significant at the 2 per cent level). These findings suggest either that the groups were different in the beginning or that the stress the trapped miners had endured was less damaging to the abilities involved in the tests.

TABLE 13

Comparison of Mean Intelligence and Personality Test Scores for Trapped and Nontrapped Miners

Test	Trapped miners (N = 19)		Nontrapped miners (N = 12)		t	p
	N	Mean	N	Mean		
Age	19	38.9	12	40.2	0.62	ns
Education	19	6.8	12	6.6	0.19	ns
Vocabulary	17	7.7	3	7.0	0.39	ns
Digit efficiency	19	8.0	12	7.7	0.28	ns
Block design	19	7.1	12	5.1	2.44[a]	.05
100 by 7's	8	90.4"	6	77.8"	0.58	ns
20 to 1	18	19.4"	12	16.0"	0.72	ns
IQ (DS & BD)	19	87.5	12	78.9	1.85	ns
Psychiatric ego strength	19	53.4	12	48.5	1.57	ns
Rorschach control	19	15.4	12	15.1	0.12	ns
Rorschach number of responses	19	7.8	12	7.2	0.42	ns
Rorschach disorganization	19	5.7	12	7.8	1.88	ns
Bender impairment	19	3.2	12	5.8	2.53	.05

[a] Corrected for inequality of variance.

The hypothesis that the trapped and nontrapped groups of miners were dissimilar samples is supported when each group is subdivided into younger and older men (Table 14). Mean scores of the younger trapped miners were significantly better than those of the older trapped miners on Block Design and Bender-Gestalt (both at the 5 per cent level, corrected for differences in variance), and the mean prorated IQ of the younger trapped miners was 13 points

TABLE 14

Mean Psychological Test Scores
of Younger and Older Trapped Miners,
and Younger and Older Nontrapped Miners

Test	Trapped miners		Nontrapped miners	
	Younger[a] (N = 8)	Older[b] (N = 8)	Younger[c] (N = 6)	Older[d] (N = 6)
Education	7.4	6.5	6.8	6.3
Vocabulary	8.5	7.3	---e	---f
Digit efficiency	9.2	7.0	7.0	8.3
Block design	9.1	6.0*	4.7	5.5
100 by 7's	77.8"	102.5"	85"	71"
20 to 1	11.7"	24.3"*	21.2"	10.8"
IQ (DS & BD)	96	82.9	76.7	81.2
Psychiatric ego strength	54.5	48.1	43.2	53.8*
Rorschach control	17.2	12.6	13.5	16.7
Rorschach number of responses	8.8	7.1	8.0	8.2
Rorschach disorganization	5.9	6.4	5.2	4.3
Bender impairment	1.75	4.4*	6.3	5.2

[a]22-35 years. [b]42-58 years. [c]29-41 years. [d]42-50 years.
[e]Only 2 Ss. [f]Only 3 Ss.
*Significantly different from younger group at the .05 level (t tests)

higher than that of the older trapped miners (not significant). Moreover, the older trapped miners' mean scores were poorer than those of their younger mates on all 11 tests and there was a tendency toward greater variability among the older men [significant for Bender-Gesalt (F ratio = 5.250, p between 10 per cent and 2 per cent)]. One inference from these results is that the older subgroup was deviant for some reason; indeed, the older subgroup did include three of the four injured miners and a fourth man whose test results suggested brain damage. However, the test scores of these injured men did not differ appreciably from the other older miners nor were they greatly different from the scores of the older and younger nontrapped miners. The likeness of the scores of the older trapped group and the older and younger nontrapped groups, in contrast to the higher scores of the younger trapped miners, suggests that the younger trapped miners were uniquely proficient on the tests. The conclusion is that the proficiency of the younger members of the trapped group contributed to the finding that trapped miners as a whole did better than nontrapped miners. Nevertheless, it is notable that a period of entrapment with attendant physiological and sensory deprivation plus extreme psychological threat apparently had no more effect on test scores than the experience that nontrapped miners had endured.

Age (and probably years of experience) may have had differential effects on the trapped and nontrapped miners. In the trapped group, younger miners' mean scores were better than their older mates on all 11 tests; in the nontrapped group, older miners did better than younger men on all 11 tests. The probable uniqueness of the sample of younger trapped miners, mentioned above, may have contributed to the former differences, but the consistent reversal suggests that age made miners more susceptible to the stress of entrapment while it made for resistance to the stress of rescue work and living above ground during the emergency.

The Sentence Completion responses of the two groups of miners revealed differing attitudes and thought processes. The trapped miners' responses expressive of fear, anxiety, and needs were focused on themselves while those of the nontrapped group were more concerned with external factors such as family and economic problems. The anxiety of the trapped subjects did have one external focus: more of them expressed an aversion to the mine.

The trapped men were more open in expressing hostility toward bosses while the nontrapped were more positive about bosses and superiors. Responses to other items indicated that the mental and emotional condition of the men who had been trapped made it difficult for them to think clearly and make up their minds on various issues. Thus more of them were evasive and noncommital to items about leaders, relatives, fears, the mine, workmates, and the future; and more expressed anxious anticipation to the item about leaving Minetown. By contrast, more of the nontrapped miners committed themselves on such problems, giving definitely positive or negative statements; more of them gave positive evaluative responses about workmates, relatives, and their minds and nerves. The nontrapped miners' positive evaluation of their minds and nerves suggests that their self-esteem was more intact than that of the trapped miners. More of the nontrapped miners seemed to be facing up to the facts on all sides. They were both more pessimistic about the future and less willing to leave Minetown; they were apparently aware of the unhappy consequences of the mine closing, but at the same time they did not want to leave their established homes and start all over again in a new setting.

Intertest Correlations Within Groups

As a means of throwing further light on the differences between the two groups of miners, rank order correlations (corrected for ties) were computed between all test scores within each group (Tables 15 and 16). Of the 121 correlations, 32 were significant at the 5 per cent level or better, whereas only 6 might be expected by chance. The significant correlations were, therefore, taken as pointing up hypotheses.

Education was positively associated with intellectual efficiency in the trapped miners (as measured by the prorated IQ), while such a relationship did not hold in the nontrapped group. Since education is normally correlated with intelligence (Anastasi, 1958), it may be inferred that the nontrapped miners' emotional state disrupted the intellectual abilities measured in this study in an unsystematic fashion. Accuracy of perception and visual-motor coordination as reflected in the Bender-Gestalt drawing test gave a different picture: amount of education was associated with this ability (1 per cent level) in the nontrapped group. It may be that increased formal education made for specific overlearning in this area and hence made the skill less susceptible to impairment.

Psychiatric ratings of ego strength were associated with contrasting sets of measures in the trapped and nontrapped miners. In the trapped group, they were correlated significantly with vocabulary (5 per cent level) and with Rorschach control (2 per cent level), while they tended to correlate positively with Rorschach number of responses and negatively with Rorschach disorganization. In the nontrapped group, psychiatric ego strength was not correlated with Rorschach measures, but showed a general tendency to be positively related to intellectual efficiency, including Block Design, Digit Efficiency (5 per cent level), IQ, Bender-Gestalt proficiency, and education. The meaning of these differing relationships is not immediately clear. One possibility is that psychiatric ego strength ratings were based on two different qualities in the two groups: on evidence of the individual's general ability as exhibited in education and performance record in the nontrapped group and on the individual's reaction to the stress of entrapment and ability to communicate in the trapped group. Insofar as the former would probably depend to a considerable extent on intelligence, while the latter was more a function of emotional make-up, psychiatric judgment was apparently a valid measuring instrument in that it selected and was associated with the variables that were relevant to the situation. Further, in line with other test results, psychiatric ratings favored the younger and less experienced men in the trapped group and the older and more experienced men in the nontrapped group.

TABLE 15

Correlation Matrix for Trapped Miners (N = 19)

	Education	Psychiatric ego strength	Rorschach control	Rorschach number of responses	Rorschach disorganization	B-G errors	WAIS IQ	Block design	Digit efficiency	Years experience	Vocabulary
Age	-.26	-.32	-.45	-.29	.23	.55	-.29	-.53	-.44	.86	-.18
Education		.23	.20	.28	-.09	-.22	.47	.41	.39	-.53	.79
Psychiatric ego strength			.58	.45	-.40	-.01	-.10	.06	-.08	-.51	.49
Rorschach control				.85	.04	-.33	.33	.27	.34	.66	.53
Rorschach number of responses					.22	-.02	-.08	.08	-.25	-.57	.24
Rorschach disorganization						.01	-.20	.25	-.24	.15	-.39
B-G errors							-.47	-.56	-.51	.46	-.09
WAIS IQ								.85	.88	-.35	.41
Block design									.71	-.60	.37
Digit efficiency										-.37	.40
Years experience											-.57

Note: Rho of .46 significant at 5 per cent level (two-tailed); rho of .55 significant at 2 per cent level. Figures are rank order correlations corrected for ties.

TABLE 16

Correlation Matrix for Nontrapped Miners (N = 12)

	Education	Psychiatric ego strength	Rorschach control	Rorschach number of responses	Rorschach disorgani- zation	B-G errors	WAIS IQ	Block design	Digit effi- ciency	Years expe- rience
Age	.16	.58	.20	-.18	-.17	-.16	.37	.15	.63	.88
Education		.49	.34	-.13	-.32	-.82	.11	.07	.33	.15
Psychiatric ego strength			.47	.21	-.05	-.47	.56	.48	.58	.56
Rorschach con- trol				.67	-.15	-.28	.11	.46	.16	.18
Rorschach number of responses					.56	-.19	-.13	.37	-.13	-.28
Rorschach dis- organization						.27	-.41	-.25	-.61	-.28
B-G errors							-.07	-.42	-.06	-.22
WAIS IQ								.66	.82	.48
Block design									.47	.27
Digit efficiency										.62

Note: Rho of .58 significant at 5 per cent level (two-tailed); rho of .71 significant at 2 per cent level. Figures are rank order correlations corrected for ties.

There was a general tendency for age and years of experience in mining to be correlated in contrasting ways with intelligence and personality measures in the two groups. In the trapped group, years experience and age had negative relationships with all measures of intelligence and personality (8 of the 20 were significant). Most of the correlations were higher for years experience than for age, and partial correlations between Rorschach control, years experience, and age, revealed that years experience was the significant variable (r_p = -.60, significant at the 2 per cent level) for this measure. In the nontrapped group, on the other hand, age and years experience had positive relationships with all but two of the measures of intelligence and personality. This reversal of trends between the trapped and nontrapped miners is consistent with group differences already noted in Table 16. It supports the hypothesis that age was a liability for trapped miners and an asset for nontrapped miners. In the trapped group, Vocabulary and Rorschach number of responses were negatively correlated with years experience (2 per cent level), indicating that years experience together with being trapped had led to some constriction or repression of thought and imagination. This suggests the presence of a depressive element in the reaction of the more experienced trapped miners. This inference is supported by the Sentence Completion finding that older trapped miners expressed more self-deprecation than their younger mates. Just why years of living and working with danger should cause this type of reaction is not clear. It may be that recurring accidents in the mines together with their own and their wives' reservations about the occupation had sometimes made them feel they should quit. Under these conditions, despondency and sense of hopelessness and guilt could well be their reaction.

Length of Entrapment and Test Results

The test results of the trapped group of twelve, entrapped for 6 1/2 days (six-day group), and the trapped group of six plus the semi-isolated miner, entrapped for 8 1/2 days (eight-day group), were compared (Table 17). Their mean age and education did not differ. The eight-day group exhibited more variability in number of responses to the Rorschach (F ratio = 4.326, 5 per cent level) and showed a near-significant tendency to give more responses ($\underline{t}$ was 2.167, slightly lower than that required for significance with unequal variances). In addition, these miners got higher Rorschach disorganization scores than the six-day group. One interpretation of these findings is that the members of the eight-day group were experiencing a post-crisis excitement or euphoria (Wallace, 1956) that involved an exaggerated press to talk without adequate care for what was said. Why it should be particularly evident in the eight-day group is another question. One hypothesis is that, being trapped two days longer and being less certain of rescue, they were closer to hopeless despair and rescue produced an excitement reaction. A second hypothesis is that being a smaller group they were subject to more social control (as indicated by the total initiation scores of the small group, in Chapter 5), and rescue led to a release of impulses and emotion. The findings might also be interpreted as an expression of exaggerated dependency needs. This inference is supported by the eight-day group's apparently stronger dependency needs as found in the Sentence Completion test. More of them made positive evaluative statements about relatives, superiors, leaders, religious people, God, and home, indicating a strong need to respond positively to others, which usually involves a tendency to please others by responding to their expectations. This may have led them to keep responding to the Rorschach examiner. Any or all of these three factors may have contributed to the eight-day group's Rorschach performance.

TABLE 17

Mean Intelligence and Personality Test Scores of Eight-Day Group and Six-Day Group

Test	Eight-day group (N = 7)		Six-day group (N = 12)		t	p
	N	Mean	N	Mean		
Age	7	39.0	12	38.8	0.32	ns
Education	7	6.6	12	6.9	0.41	ns
Vocabulary	6	6.8	11	8.2	0.96	ns
Digit efficiency	7	8.0	12	8.0	--	ns
Block design	7	6.6	12	7.4	0.43	ns
100 by 7's	2	71"	6	97"	1.14	ns
20 to 1	7	19.4"	12	19.1"	0.23	ns
IQ (DS & BD)	7	88.9	12	88.8	0.31	ns
Psychiatric ego strength	7	55.8	12	52.9	0.89	ns
Rorschach control	7	18.7	12	13.8	1.22	ns
Rorschach number of responses	7	10.3	12	6.4	2.17[a]	ns
Rorschach disorganization	7	8.2	12	4.3	2.83	.02
Bender impairment	7	3.7	12	2.9	1.33	ns

[a]Corrected for inequality of variance.

The Sentence Completion data brought out other differences between the eight-day group and the six-day group. The eight-day group was more self-oriented in their expressed needs and worries while more of the six-day group focused on external issues such as their economic problems. This tendency of the eight-day group to emphasize personal and subjective concern was even reflected in their responses to the item, "When I am afraid..." They tended to give responses descriptive of their own reactions whereas more of the six-day group said they would seek help from God or other sources outside themselves. Apparently, the men in the eight-day group were less able to look beyond themselves.

The responses of the eight-day group to a number of the items indicated that they were concerned to cast themselves in a good light and to deny their feelings. Whereas more of the six-day group gave self-depreciation responses and stated that their greatest weakness was lack of emotional control, more of the eight-day group were evasive about their greatest weakness, said their "nerves" were good, and attributed socially valued characteristics to themselves. These attitudes of the eight-day group might be related to the hypothesized excitement they experienced after rescue. However, another explanation is perhaps more reasonable. The harrowing experience of living with the moaning and pleading pinned miner for five days and doing virtually nothing to save him may have made them defensive about their personal adequacy.

The six-day group tended to be evasive about the future and about the possibility of leaving Minetown. Apparently, they were in a state of conflict and of uncertainty, with their feelings and reality considerations about equally balanced on these matters. By contrast, nearly one-half of the eight-day group were quite positive and optimistic about the future, again reflecting a rather unrealistic tendency to make positive statements about things. On the other hand, more of them were negative about leaving Minetown. It is inferred that both of these attitudes reflect their strong dependency needs and some lack of consideration for the facts of their situation. Assuredly, strongly activated dependency needs would not incline them to think of breaking their ties with their homes, friends, and familiar surroundings in Minetown. Granting these inferences, the expressed attitudes of the eight-day group seemed to be more determined by subjective feelings than by external facts.

The Effects of Entrapment

Contrary to expectations, the trapped miners did not score lower on psychological tests than did the nontrapped miners. Rather, they showed some tendency to do better in spite of the fact that the trapped group included several injured. The probability that the younger members of the trapped group were an unique sample with more ability and resources did not adequately account for the lack of effects from entrapment. The fact that testing was done from 5 to 23 days after rescue, thus giving trapped miners time to recover, may have contributed to the results. Again, however, this would not seem to be a complete explanation, for the interviewer noted that 15 of the 19 miners were tense and anxious (Appendix B), and the Sentence Completion test indicated anxiety.

An hypothesis more in keeping with all the facts is that the nontrapped miners did not constitute a control group that was free from stress. Rather, the two groups had been subject to stresses that differed in kind, degree, and the manner in which they were handled. All the nontrapped miners, although they were above ground, were faced with a number of anxiety-producing situations and prospects friends and relatives had been lost, the whole community was in a state of anxious anticipation, and they faced the prospect of the last mine in Minetown closing. Moreover, two-thirds of them were engaged in the extremely difficult, dangerous, and nauseating rescue work. On the other hand, they were not deprived of external stimulation and were relatively free to make choices and decisions, to carry out goal-directed activities, and to express their emotions and impulses. By contrast, the trapped miners were helpless and faced the threat of death. For much of their time underground, they did not have the same free alternatives for imagining, anticipating, choosing, and doing. At the same time, they were hungry, thirsty, bruised, and uncomfortable. Their condition and situation would certainly arouse anxiety. During entrapment, however, the group social controls tended to suppress overt expressions of anxiety in the trapped groups. Thus overt and expressive reactions to the situation were controlled, while autonomic reactions and feeling states were probably accentuated. The combination of autonomic and subjective anxiety on the one hand and control of expression on the other could well account for the trapped miners' greater self-preoccupation and indecision on the Sentence Completion test and for their relatively well-controlled performance on other tests. The nontrapped miners' circumstances, however, would probably orient them to external problems and facilitate expressed decisions, as reflected in their Sentence Completion responses, and make for less overt control of behavior, as reflected in other tests.

The anxiety reaction in the trapped miners had several depressive features self-preoccupation, self-depreciation, indecision, and constriction of thought and imagination. Wolfenstein (1957) has noted that disaster victims may have depressive reactions and tend to focus their emotional energies on themselves. The anxiety reaction of the nontrapped miners was more overt and situational. The suggestion that age and years experience made for poorer test performance in the trapped group suggests that the depressive anxiety, which is more likely in older people (Mayer-Gross, Slater, & Roth, 1955), had increased to the point where it affected test results adversely (Rapaport et al., 1946). Older and more experienced members of the nontrapped group, on the other hand, were probably more adapted to their situation and hence had handled it better than the younger, less experienced men.

The eight-day group exhibited more subjective anxieties, stronger dependency needs, and a tendency for members to cast themselves in a good light, but members were not more impaired on intellectual and performance tasks. They did manifest less perceptual control on the Rorschach and a tendency to give more responses. This may have been due to their longer confinement. Other conditions, however, may also have contributed to these tendencies, including the small size of the group with its tighter social controls and the presence of the pinned miner. Thus length of entrapment did not involve simply an extension of time, but was complicated by a number of social and situational factors, all of which presumably affected the psychological findings.

CHAPTER 7

THE RELATIONSHIP BETWEEN PSYCHOLOGICAL DATA
AND INITIATIONS

In this chapter, the trapped miners' initiations while entrapped are related to their psychological test scores. An initiation was defined as an act that originates an extended sequence of behavior. Initiations were used as the basis for three types of scores: <u>Initiation Perception</u>, the number of initiations a subject reported having observed, <u>"I" Initiation</u>, the number of initiations the subject attributed to himself, <u>Group Evaluation</u>, the number of initiations that all the other members of the group attributed to the subject. As the only measure of behavior while entrapped, they were correlated (rank order) with test scores, as well as with education, psychiatric ratings of ego strength, and years of mining experience.

Taking the 5 per cent level as the criterion significance level, 12 of the 99 correlation coefficients were significant (Table 18). Only 5 criterion correlations could be expected by chance, the reported relationships and interpretative statements should be taken as hypotheses

Relations among Initiation Measures

Initiation Perception correlated significantly with "I" Initiation (Table 18). In other words, the more initiations a man reported, the more he attributed to himself. However, examination of the correlations for the separate groups by periods revealed that the relationship was largely a function of the high correlations for the six-day group (at the 2 per cent level in each period). The correlations were in the same direction for the eight-day group, but were not significant. Two subjects, C6 and D6, contributed the greater part of the variance from the regression line in the eight-day group. C6 reported a large number of initiation perceptions and credited himself with very few. This man was 49 years old, had been a miner for 32 years, and had suffered two major accidents in recent years. He was partially deaf, and his mates said that he became completely deaf during entrapment. This cannot be true because he reported events that he could have perceived only by auditory means. However, this miner was the focus of considerable teasing and irritation while entrapped. Assuming that he accepted the role of the "butt," it seems possible that he did not take himself seriously and so reported few "I" initiations. It is also possible that his low intellectual capacity (prorated IQ of 54) and apparent brain damage contributed to the inconsistency between his "I" Initiation and Initiation Perception scores.

TABLE 18

Rank Order Correlations Between Test Scores
and the Three Measures of Initiation for Escape and
Survival Periods Separately and Combined (N = 18)

Test	Initiation perceptions			"I" initiations			Group evaluation		
	Escape	Survival	Combined	Escape	Survival	Combined	Escape	Survival	Combined
Years experience	-.46	-.33	-.34	-.35	-.28	-.31	-.09	-.29	-.19
Psychiatric ego strength	.38	.64	.52	.35	.63	.57	-.34	.50	.07
Rorschach control	-.01	.21	.21	.18	.24	.37	-.02	.43	.31
Rorschach number of responses	-.01	.09	-.04	.18	.22	.25	.24	.24	.18
Rorschach disorganization	-.37	-.15	-.39	-.08	.04	-.12	.21	-.20	.25
B-G errors	-.36	-.02	-.23	-.31	-.08	-.16	-.23	-.37	-.40
WAIS IQ	.21	-.18	.08	.41	-.03	.22	.16	.44	.38
Block design	.41	-.02	.28	.57	.07	.35	.35	.60	.52
Digit efficiency	.24	-.16	.13	.21	-.05	.16	.15	.24	.10
Vocabulary	.32	.46	.42	.29	.39	.39	-.46	.37	.05
Initiation perception[a]	-	-	-	.68	.83	.70	.10	.17	.17
"I" initiation[a]	-	-	-	-	-	-	.24	.43	.50

Note: Rho of .47 significant at 5 per cent level (two-tailed).

[a]Correlated within periods.

The other miner who reduced the correlation between Initiation Perception and "I" Initiation was D6. He reported only a moderate number of initiation perceptions and credited himself with nearly all of them. In the escape period, his Initiation Perception score was the lowest in the group, and this was in keeping with his relative inactivity in this period--even by his own story. It seemed that he perceived his role as that of consultant and adviser, directing important moves from the central location from which the others went out to search for means of escape. In the survival period, the tendency to see himself as the central figure became more noticeable as he reported nine initiation perceptions and attributed all of them to himself. In this instance, however, his perception of his role was well confirmed by the reports of his mates· of 15 initiations attributed to various members of the group by all the rest, D6 received credit for 12, meaning that his mates saw him as responsible for most of the significant initiations in the survival period. It was clear that this miner was recognized as the survival leader in the eight-day group. However, his perception and recall of events was very atypical· restricted and selective in a way to give himself maximum credit. Presumably he had a strong need to be ascendent and to be recognized by the group. The personal characteristics, including his Negro ancestry, which may account for this are discussed in Chapter 8.

For the escape and survival periods combined, there was a positive relationship between the number of initiations the group attributed to a man and the number he credited to himself. Again, the six-day group was mainly responsible for this correlation, and primarily in the survival period. However, it is noteworthy that when the three men are omitted whose "I" Initiation and Group Evaluation were most discrepant, the rho for the rest of the total group was increased from .50 to .75. Of these three, F6 and O12 rated themselves high on initiations and received virtually no credit from the group, while G12 rated himself low (his interview was not recorded) but was credited with many initiations by the group. G12 had been treated for a depressive reaction about two years before the disaster. O12 was injured and immobile during entombment, and he exhibited paranoid tendencies, which may have led to his apparent need to give himself considerable credit in the situation. F6 was apparently the most unstable man in the group, and he "thought" he was injured by the bump and remained virtually immobile for the 8 1/2 days.

Psychological Tests and Initiations

Only one test, the Block Design subtest, was significantly related to Group Evaluation for the escape and survival periods together. Apparently, analytic thinking and the ability underlying form perception were important qualities for general leadership during entrapment. The test with the next highest correlation with group evaluation was the Bender-Gestalt drawing test (10 per cent level). It is noteworthy that the primary source of the correlations for these two tests was in the survival period. Block Design was also associated with "I" Initiation for the escape period, indicating that ability on this test was related to self-reports of participation in escape attempts.

Some individual tests were not significantly correlated with initiations, apparently because of the atypical scores of one or two men. For example, the rho between Rorschach control and Group Evaluation for escape and survival combined was .31; without miner G12, it was .55. G12 was first in initiations according to his mates, but he had the poorest Rorschach, also, he had previously been treated for a depressive reaction. Such findings indicate that tests may be related to most men's initiation behavior in situations like that of entrapment underground, but that the relationship may break down with a few unusual individuals.

Fifteen of the 18 correlations of psychological tests with "I" Initiations for the escape and survival period separately were in the positive direction (Table 18), and there was no difference in this regard between periods. Assuming that chance would result in an equal number of positive and negative correlations, this trend toward positive correlations is very significant (chi square = 6.722, p = 1 per cent). This indicates that there was a general tendency for the trapped miners with the "better" test results to attribute more initiations to themselves. In view of the previously mentioned correlation between "I" Initiation and Group Evaluation, it is suggested that men with better test performance did initiate more action. This inference is supported by the finding of a similar trend toward positive correlations between tests and Group Evaluation-- 14 of the 18 rho's were positive (chi square = 4.500, p = 5 per cent). Taking Group Evaluation as a measure of the group's perception of an individual's contribution to its current goals, there was a trend relationship between proficiency on psychological tests and leadership.

Psychiatric ratings of ego strength were significantly correlated with "I" Initiation and Initiation Perception for the escape and survival periods combined. Apparently psychiatric judgment was influenced by the individual's tendency to give a detailed account of events and his tendency to give himself credit for initiation behavior. It is notable that the main source of these correlations was in the survival period, suggesting that the traits subsumed under the term "ego strength" were judged to be characteristic of leaders in the survival period but not of the leaders in the escape period. Indeed, a negative correlation of -.34 between psychiatric ego strength and

Group Evaluation in escape, in contrast to a positive correlation of .42 (10 per cent level) in survival, indicates that ego strength traits as judged by psychiatrists may be associated with leadership as measured in one kind of situation, but may contraindicate leadership in another kind of situation That this selectivity was not confined to psychiatry is indicated by the correlations between survival Group Evaluation and tests being consistently higher than those between escape Group Evaluation and tests. This suggests that the personal qualities associated with leadership in the survival period are more akin to the qualities that meet everyday criteria of achievement and adjustment in our society.

The essential contrast between the roles required in the escape and the survival periods is reflected in the lack of correlation between Group Evaluation for the two periods (rho = .16), whereas the correlations between periods were .50 for Initiation Perception (5 per cent level) and .53 for "I" Initiation (5 per cent level). Each miner tended to see himself as either consistently active or passive during the whole period of entrapment, but he did not find his colleagues so consistent. Some he saw as leaders during the escape period, some as leaders during the survival period. In general, the perceived escape leaders were not the same men as the perceived survival leaders the individuals whom they perceived as promoting their goals differed for the two periods.

One trend highlighted the contrasting qualities of the perceived leaders for the escape and survival periods. Group Evaluation in survival correlated .37 with Vocabulary and .51 (5 per cent level) with education. For the escape period, Group Evaluation was correlated -.46 (10 per cent level) with Vocabulary and -.32 with education. Apparently, training and proficiency with words were assets in promoting group goals in survival, while these skills tended to have negative value when the goals were focused on escape attempts. It is suggested that the differing circumstances of survival and escape periods precipitated the emergence of intellectualizers and task-oriented individuals as leaders in their respective periods.

To check some of the relationships between leadership and personal qualities more closely, the test scores of selected miners were examined. One apparent relationship was that between performance tasks and Group Evaluation in the escape period. Of the four top initiators in this period (G12, A6, M12, E6), all obtained higher scores on Block Design than on Vocabulary. On the other hand, three of the four highest initiators in the survival period (K12, N12, D6), tended to be "intellectualizers," as indicated by their high Vocabulary scores and the expressive quality of their sentence completions; two were also proficient on Digit Span and on the two other counting tasks. The fourth high initiator in survival, Q12, was an exception on verbal material. Thus, the findings indicate high performance abilities apparently characterized initiators in the escape period, while verbal-intellectual abilities were predominant only in the high initiators of the survival period.

Some Rorschach test results were related to behavioral criteria. Most of the outstanding men for Group Evaluation in the survival period had good or very good Rorschach control scores (K12, D6, Q12). The other high initiator in this period, N12, had a rather constricted record but showed less than average Rorschach disorganization. A "Rorschach composite" score was computed by subtracting Rorschach disorganization from Rorschach control. Of the six highest and four lowest initiators (group evaluation, both periods), all of them uninjured miners, the Rorschach composite score correctly differentiated eight of the ten. This contrasted with two correctly differentiated among the ten by an independent clinical appraisal of the Rorschach protocols. It was clear that quantified Rorschach data were quite efficient in identifying leaders and those who contributed least, although it did not discriminate well in the middle of the distribution. The quantified Rorschach was also much more successful than a clinical appraisal of those data.

Examination of the Block Design and Bender-Gestalt scores of the same ten high and low initiators revealed that each of these tests differentiated the two levels of leadership with but one error.

The three subjects (C6, O12, Q12) who attributed an unusually large number of initiations to "they" gave no Human or Texture responses to the ink blots. Only G12, the subject who had had a depressive breakdown two years previously, gave a similar Rorschach in this respect. The sense of apartness implied in "they" initiations thus may be associated with lack of dependency needs and lack of empathy as measured by the Rorschach. Taking the incidence of Human responses alone, six subjects gave two or more (J12, K12, B6, N12, D6, P12). Of this group, it was inferred that J12 tended to use fantasy as an escape and defense. Of the others, K12, N12, and D6 were top initiators in the survival period; P12, though seriously injured and immobile, was credited with more initiations than nine other trapped men in the escape period; and B6 was second highest initiator in his group in the survival period and third highest in the escape period. On the other hand, Q12, the second highest initiator in the survival period in the group of twelve, gave no Human responses. Nevertheless, the production of Human responses would seem to be associated with initiation behavior.

The influence of mining experience on a miner's initiation behavior turned out to be indeterminate. Although there was a consistent tendency for years experience to be negatively correlated with measures of initiation (Table 18), this trend was probably the result of some interviews being untaped and some men being injured. Of the four miners with untaped interviews, two were senior in mining experience and one was average. Of the three injured, two were senior in mining experience and one was average. Thus, any effects of mining experience on initiations were confounded by other variables and therefore indeterminable.

The Six-Day and Eight-Day Groups

The group of twelve and the group of six were trapped separately for 6 1/2 and 8 1/2 days respectively. Groups of this size are too small for detailed comparison but trends may be suggestive. Of the 20 correlations between Group Evaluation and test scores, 13 were positive (in terms of proficiency) for the six-day group and 15 were positive for the eight-day group. Of the 20 rank correlations between tests and "I" Initiation, 15 were positive for the six-day group and 18 for the eight-day group. However, of the 20 correlations between Initiation Perception and test scores, all 20 were positive for the six-day group while only 6 were positive for the eight-day group.

It will be remembered that the eight-day group not only had a longer period of entrapment but also more rigid group social control, and in addition it had the harrowing incident of the pinned miner. Presumably, these three factors affected the individuals differently: those with better test performance reported fewer initiations and those who reported more initiations had poorer test performances. It is only in this way that the negative correlations of the eight-day group can be explained. Independent data were insufficient to justify an interpretation of this somewhat negative relationship between test proficiency and Initiation Perception.

Discussion

The findings in this chapter do not constitute a satisfactory basis for generalizations or predictions. Sampling problems impose this reservation. Moreover, all of the tests were given after the disaster, when test results would presumably be reflecting some combination of individual qualities and the effects of stress. To some extent, the study tells more about the tests than it does about reactions to stress. Nevertheless, the findings have raised a number of interesting issues.

The general tendency for test proficiency to be related to initiation behavior, either as perceived by the group or by the individual himself, is not surprising. It suggests that initiation has some relationship with the criteria for test performance. Why the relationship was rather general and not too strong is an important question. The answer would seem to involve three areas the psychological tests, the individuals examined, and measures of initiation.

The psychological tests used in the study were probably not the most appropriate. The Block Design subtest was perhaps best in terms of its relationship with initiations and other tests. This is consistent with its record in intelligence and personality studies (Rapaport et al., 1946, Wechsler, 1955). The Bender-Gestalt drawing test was fairly discriminative. Its usefulness might well be augmented by selected changes in scoring procedure. Whether its discriminative power would be evident in a prestress situation is another question. The Vocabulary subtest exhibited selective relationships. This, together with its general stability, would seem to recommend some such measure for similar studies. The Digit Span subtest appeared to contribute little to the results.

The use of the Rorschach as a research instrument was not vindicated in the present study. Nevertheless, a simple method of quantifying the productions would seem to be indicated in preference to a clinical appraisal. This is not to say that a clinician could not do as well if he knew what to look for. The point is that what he looks for is determined by his own personality, experience, training, and definition of the situation. To use clinical appraisals is to introduce variables without identifying and controlling them. The Rorschach may well suffer by the vagueness of the stimuli it presents. Predictable behavior does not occur in a vacuum but in response to relatively specific situations.

While the Sentence Completion test pointed up some differential attitudes in the groups of miners, it was an inefficient instrument: not only may something be lost when a subject can pause to write answers, but undue time is consumed by the very act of writing. Moreover, much of the information cannot be used when the responses of small groups of subjects are categorized. A multiple-choice or true-false questionnaire would probably have produced more data that are pertinent and quantifiable. In general, the use of empirically well-established psychological tests would seem to be indicated. Even if they did not relate to measures of behavior, at least interpretation of standardized tests would be less difficult.

The instances in which one or two individuals deviant on a particular test "spoiled" a probable relationship between test results and initiations points up the role of individual differences. It is noteworthy that several tests, including Block Design, Bender-Gestalt, and Rorschach composite, were efficient in differentiating the miners at either end of the initiation scale. The same tests were less successful in discriminating the rank order of these individuals and in differentiating miners in the middle of the initiation scale. The basic problem may be stated as an ignorance of the relevant variables or a failure to measure and relate them.

The degree of relationship between test performance and some overt behavior will also depend on the nature of that overt behavior. The initiations of the miners were the only measure of their behavior while trapped. Such behavior is not simple. Operationally, it was a respondent's oral report of a remembered verbal or physical act that a researcher judged as leading to an extended sequence of behavior. Reported initiation would presumably be determined in a complex way by perception, memory, emotional factors, and the respondent's perception of and reaction to the interviewer and the question. In the light of these influences, it is remarkable that a separate measure of behavior, psychological tests, should be related to initiations at all.

It would have been preferable to have had more than one interview with every miner. This would have permitted determination of the reliability of the respondent's account, as well as an analysis beyond initiations in terms of the dynamics of interaction. Different types of interaction might also manifest more specific relationships with psychological test performance.

CHAPTER 8

PSYCHOLOGICAL AND BEHAVIORAL ANALYSIS OF SELECTED INDIVIDUALS

Individual appraisals of a few of the miners who played distinctive roles during entrapment will be given in some detail: E6 and G12, the escape leaders, M12, a subleader during escape; D6 and K12, the survival leaders; Q12, a willing follower, J12, a passive-dependent man, F6, a disorganized man; and X6, the isolated man.

Miner X6 is discussed because of the unusual array of conditions, characteristics, and behavior involved. The other eight miners were selected either because of the unusual nature of their case or because of the special role they played in their particular group. It seemed useful to make a more detailed examination of the individual as a person in contrast with the examination of him as an impersonal member of a group. The other miners were not less interesting but their histories presented little different material.

E6: An Escape Leader

E6, a datal miner with four years experience, responded directly and realistically after the bump. During the remainder of the escape period he was active in exploring all possible ways out, usually with his younger friend, B6. E6 had the highest Group Evaluation during the escape period, but when the group of six gave up their attempts to escape, E6 became relatively inactive and acknowledged the ascendancy of D6. He was ranked near the bottom in group evaluation for the survival period. Although it was E6 who suggested that they drink urine and eat coal and who was still bent on escape as late as the sixth day, he was apparently not the kind of man who could give emotional support and raise morale during the long wait. He was the only man who said he was afraid of being blamed for the treatment of the pinned miner. E6's account of the entrapment was the most detailed statement obtained from the group of six he reported almost twice as many initiations as any of his companions.

This escape leader was 35 years old. He was raised on a farm and had to do regular farm work at a very early age. When 12 years old, he left school in grade 5 to help his ailing father, who died two years later. He was very grieved by his father's death. He worked in the woods and on the railway before becoming a miner. Married at 19, he had four children, the youngest being microcephalic. E6 shouldered a man's responsibility early in life and handled recurrent stresses in a reasonably adaptive manner.

He was in fairly good condition when rescued. His hospitalization lasted only the minimum of two days. The major complaints he reported were headaches, cramps, weakness, and restlessness. However, after the first night or two he slept well, not dreaming. When interviewed ten days after rescue, he appeared rather tense and smoked heavily. He expressed hostility toward the company and "all those damned reports (news accounts of the trapped man's experience), they are getting everything all mixed up." He was rather depressed and cynical about the future, but he stated that he would go back to the mines rather than let his wife and children starve.

When tested ten days after rescue, E6 obtained a prorated IQ of 79. His wide range of interests and above-average productive energy would suggest that his intellectual potential was higher, and probably that the impairment reflected in his IQ score was due to the emotional strain of his experience. He displayed considerably better than average Rorschach control, and on more simple performance tasks he was better than average for the trapped miners, indicating that practical judgment and planning were relatively intact. The test data revealed that he was most efficient on practical performance tasks rather than on problems requiring intellectual skills. Moreover, his reality testing and organization improved under stress. Emotional responsiveness was not covered up as much as in some of the miners, but it was hostility that was expressed. This trend to express his feelings was also seen in a higher than average anxious concern about the future and getting a job.

G12. An Escape Leader

In the group of twelve, G12 was the primary leader during the escape period. His performance was especially outstanding in helping his eleven fellow survivors in the immediate post-impact period when most were dazed and partly buried. He himself received no serious injury in the bump, and although he said he felt "turned upside down, both inside and outside," he was immediately able to help his companions from the rubble. An uninjured but dazed miner said, "Of course, G12 went by me to help the other boys out."

During the remainder of the escape period G12 shared leadership of the group with M12. G12's role was not so much to lead escape attempts as to act as the chief adviser for them. His companions seemed to regard him as the most experienced miner. They attributed the highest initiation rank to him during this period.

G12, however, received credit as the third lowest initiator during the survival period. Like E6, he was the recognized leader of his group as long as the group's goals were to find their way out of the mine, when the goals were shifted to handling thirst, keeping up morale, and getting a message to the rescuers, G12 was not perceived as promoting these ends. He gave up hope of being rescued during the last two days of the entrapment·

> I began to kind of think what way would I be if I died; whether I would be melancholy, or whether my breathing would be irregular, or whether my heart would stop. And for a good while I thought about this, whether death would creep on me, or whether I would have any pain.

However, he kept his pessimism to himself· "At times tears came into my eyes, welled up in my eyes, but I never broke down, I never cried." He described several anxiety attacks "My heart began to pound, my breathing was difficult, my skin began to feel like a corpse." He felt that this was "due to nerves."

G12 was 47 years old and had spent 27 years in the mines. He was a contract miner until one year before the bump, when he changed to datal work because he felt he was not physically able to handle contract work. G12 reached grade 6 in school, after failing several grades. He reported exceptional shyness with girls and people in general during his early life. He was married at the age of 27 and had a 16-year-old daughter. He did not take part in community affairs nor did he visit with neighbors. He liked hunting, watching television, and raising chickens. G12 had a "nervous spell" at the age of 19, when he would not go out and refused to go to work. He went away and worked on a farm for a year, then returned to the mines. In 1954, he was given electroconvulsive therapy for a "depressive reaction." He was discharged from the hospital as "well" after two months of treatment. After being rescued he said, "I never broke. This is not so bad for a man that was two months in the (mental) hospital."

When interviewed five days after rescue, G12 had a tired, somewhat apathetic expression. He was cooperative, friendly, and helpful, but he tried to avoid talking about his experiences during entrapment. After his period of hospitalization, he reported that he felt tired, restless, and tense most of the time. He did not like being alone.

When tested six days after rescue, G12's prorated IQ was 77. Concentration was especially poor, but he showed considerable ability to persist (he managed to do the 100 backwards by 7's, though he took some time and had 3 errors) and to improve under added stress (the quality of his responses on the Rorschach improved on the inquiry). Range of ideas and interests was very restricted. His Rorschach productions were the poorest of the trapped miners, with perseveration, disorganization, inaccuracy of perception, and inappropriate emotional responsiveness. He was preoccupied with shattered bodies and unable to see the obvious. There was evidence of much latent tension. On the other hand, he was one of the two to give controlled color responses, indicating an area of emotional control.

Although G12's practical performance abilities were poorer than average, they were not as poor as those of some of the trapped miners. He expressed more fear of the mine than others and volunteered the information that he would like to move away. On the whole, G12's test results indicated extreme and disintegrating anxiety. Nevertheless, his practical performance abilities were not the worst, and he showed capacity to control emotion and to improve his performance when added demands were made on him.

In the light of this man's history and the postdisaster psychological picture it would hardly have been inferred that he was a leader during entrapment. However, G12's anxieties may well have had little outlet, for the circumstance was such that expression of uncontrolled emotion served no function and indeed was subject to group suppression. Moreover, this group apparently placed considerable value on experience, and two of the other high initiators, K12 and M12, appealed to G12's experienced opinion several times. Thus the group may have been instrumental in defining a leader's role for him. It is notable that he was perceived as leader in the escape period, when the dominant preoccupation was to explore all passages for possible escape. G12's experience would have been functional in terms of such goals. Moreover, his anxiety may have served as a drive to action. When activity was no longer emphasized, in the survival period, he had no outlet for this drive and tended to give up hope. He did not have the resources that could contribute to group morale when it was a matter of waiting, hoping, and getting a message to the rescuers.

M12   A Subleader in Escape

In the group of twelve, there was at least one subleader during the escape period. M12, a 46-year-old contract miner with nearly 28 years of mining experience, promoted escape activities with fierce persistance

> So this chap, M12, he had tired himself out completely and he'd drive us after he was down on his back. He'd get us up to try to do some job to try to get out

However, his escape activity appeared to be realistic and constructive, as is suggested by the strategy he outlined:

> (When we worked in the mine) we timbered (built packs to support the roof) it good enough that we were satisfied ourselves for our own safety. Why can't we go down and timber the waste? Then, if we get down and don't get any further, we can have a way of coming back...Well, we decided that was a good idea, so we go, we timbered it all up as good as what we could, what we figured was safe, and so got through that part of the waste (safely).

M12's Initiation Perception, "I" Initiation, and Group Evaluation scores were high for the escape period. All were low for the survival period. When escape proved impossible, M12's morale collapsed

> I thought, it's all over and we done everything we could...While I was setting there I thought about that, and it is not going to be too long before I am going to find out...for myself. I really gave up...That thought went through my mind more than often enough. I would try to brush it off.

He hoped he would be gassed rather than suffocate. The visual hallucinations that M12 reported were bizarre.

M12 had several nervous traits as a boy and was extremely self-conscious: "I couldn't go into a restaurant to eat if I thought anybody was looking at me." Before becoming a miner, he worked at various jobs, including theatre projectionist and construction laborer. At 24, he married the only girl he had ever gone out with: "I never considered marriage. No, that was the last thing in my life I ever wanted to do." M12 had been a heavy drinker during a large part of his life, and he was a heavy smoker. He was impatient, irritable, and short-tempered; he said he felt like breaking things when angry, once smashed three radios in three months. He had an unusually strong interest in hunting: "I like going to the woods, hunting. I like being alone in the woods away from people. When I'm in the woods, the world is mine." His other interest was to drink with a young miner with whom he felt close.

When rescued, M12 was in about average physical condition. When interviewed eight days after rescue, he fidgited, chain-smoked, and stood up and sat down several times. He said that after leaving the hospital and returning home, his sleep was disturbed by nightmares about the bump. "I'm as tired in the morning as when I go to bed...I am always tired." He had sudden severe headaches and had difficulty in concentrating. His hostility got out of control. "I was so cross with my children that they had to be placed with neighbors." When the children made a noise, he would scream at them and send them outside. When his three-year-old son was showing him a picture book, he became extremely agitated and threw it into the stove. He said that he felt quite guilty about this. At the same time, he did not like to be alone. He was drinking heavily and was openly hostile toward his wife; however, he said,

> When my young pal comes to the house it seems like things are different. It seems to change everything, because he has a way with me that seems to offset everything.

When tested nine days after rescue, M12 obtained a prorated IQ of 100. The primarily intellectual tests and the Bender-Gestalt drawing test were performed with very little evidence of impaired proficiency. His Rorschach productions indicated good average control with virtually no inaccuracies or evidence of disorganization. Signs of tension, emotional responsiveness, and awareness of dependency needs were completely absent on this test. In view of the stress through which this man had gone, his test results appeared to be too good and too free from anxiety effects. However, his sentence completions were rather deviant: he blocked on five items, and his responses to others indicated evasiveness, an authoritarian outlook, and rather odd thoughts. To "When I am afraid...," he wrote, "I am curious." His "greatest ambition" was "to be in the woods." His "greatest fear" and "what bothers him most" was "noise." The results on this test suggested that M12 had a thought disorder, which, in view of his very well-controlled performance on other tests, was probably paranoid in nature.

This man's suspected paranoid personality resulted in a rather unique reaction to entrapment: activeness and constructiveness during the period when there was hope and when specific goal-directed activity was in order; despair, irritability, and impaired perception or recall of events during the long wait, rather bizarre hallucinations, and virtually no impairment on most psychological tests nine days after rescue.

A number of tentative generalizations can be made about the characteristics common to the escape leaders (E6, G12, and M12)

(a) They characteristically made direct, driving attacks on problems.

(b) They perceived problems as involving physical barriers rather than interpersonal issues.

(c) They associated with one or two friends the whole group was not their frame of reference.

(d) They were rather individualistic in their expressed opinions and actions, outspoken, and aggressive.

(e) They were not particularly concerned with having the good opinion of most of the others.

(f) They lacked empathy and emotional control.

(g) Their performance abilities were better than their verbal abilities.

D6. A Survival Leader

D6 was not active in escape attempts. However, after digging his way out of the rubble (he was buried to the waist by the bump) he did go with some of the other men on the first attempt to escape. He said to the others at that time that he thought the bump had hit all three walls and would have sealed up the area. This accurate appraisal of the situation may account for, or be only a rationalization for, his lack of activity in subsequent escape attempts. He was credited by his companions as being the lowest initiator in the group for the escape period.

During the first few days, he apparently paid much attention to the pinned man; he attempted to comfort him and gave him the aspirins he happened to have with him. On about the second day, D6 suggested that the remaining water be rationed and he kept the supply, pouring out each man's portion.

When the others gave up their attempts to escape, D6 definitely became their leader. For the survival period, he was credited as being the highest initiator as well as having a large proportion of the total initiations reported. All the initiations he reported for the survival period, he attributed to himself. He suggested that they should make a noise so that the rescuers would know that they were alive. He organized shifts to beat on the pans and pipes and apparently filled in for weaker companions. The single luminous-dial watch available to the group was turned over to him.

The others said of D6:

> He was the man down there...He kept the morale up, that fellow.

> D6, he talked and joked...and then he was always singing...Kind of kept your mind.

> We had a sing-song...D6 was the leader--he was always pretty good at singing.

> That dark fellow (D6) who was with us, you know-- as soon as I heard (the rescuers) coming, I reached over and kissed him.

D6 described the attempts to keep up morale in this way:

> All the boys...seemed to be quite cooperative, as time wears on you get a little on the edgy side. Once in a while, they were kind of disgusted. I heard different things, but I figured the boys seemed to be OK...One chap there seemed to be always right shaky --you know what I mean?--and I was amazed at the way he went along. You know, before the rest of the time he was really good. I was amazed, really and truly, boy. This fellow done wonderful. I was surprised. Always laughing, I would say, "You will last for another day"...The boys would get to sleeping and I didn't know what it would be like to wake them. I

slept fairly good myself. I usually stayed awake until
I heard the boys talking and I would be doing a little
singing to myself, one thing and another, and I did a
terrible amount of praying...The boys were pretty
gallant...Everybody was really cooperative.

D6 was 46 years old. His ancestry was partially Negro, however, this was not obvious from his appearance. He had been a contract miner for 17 years and his father was a miner before him. He did not start to school until 10 years old, apparently because his mother thought he was delicate. He won a prize in grade 5 or 6 for high marks, failed grade 10 once, failed grade 11 the first time, was expelled the second year, and did not take the exams the third year.

At 28, he married a woman some 13 years younger than himself and they have had 12 children. D6 read religious literature and modern novels, played the ukelele and several other instruments, wrote popular songs, and had a tape recorder for use with his musical activities. Despite being delicate as a child, D6 displayed better than average drive, persistence, achievement, and generally adaptive behavior in his predisaster life, with an emphasis on religious and musical activities. His main associations remained within the Negro community.

D6's physical condition upon rescue was better than that of most of the others. Apart from the usual dehydration and starvation, his main complaints after rescue were restlessness and some proneness to fatigue. When interviewed nine days after rescue, D6 was friendly and cooperative. He appeared alert and intelligent although his mind sometimes wandered off the subject. After discharge from the hospital, he visited the injured men and sent sympathy cards to the widows. D6 said that he had been taking five minutes of exercises daily for about a year, to keep his weight down and to take a little off the hips and put more around the chest and shoulder, "so to make me equal." He said he would "count the nights from the twenty-third (when trapped) and...take ten minutes and would catch up" on his exercise.

D6 obtained a prorated IQ of 93 when tested ten days after rescue. There were signs to suggest that this score was lowered considerably by anxiety. He also showed some deficiency in planning and judgment. His control score on the Rorschach was the highest of any miner tested. Energy level was high, tempered with reasonably good accuracy and organization. His Rorschach disorganization score was average for the group, indicating that the stress of entrapment affected him as much as it affected others. He showed a good capacity for empathy and expressed a positive evaluation of others, but dependency needs were not in evidence. A major trait was his tendency to intellectualize rather than act. This, along with his keen awareness of his inferior status as a member of the minority Negro community in Minetown, his accurate assessment of the extent of the bump, and his faith in being rescued, may account for his inactivity in escape attempts.

D6 had the endurance and the intellectual and spiritual resources that were primary requirements during the survival period. He apparently had a strong need for attention and recognition, even to the point of being vain. His care for his companions, all of whom were white, may have satisfied a desire to have white men dependent on him. The emergent situation probably gratified these needs, giving him added strength. His activities after rescue, caring for others and raising money for the disaster fund, may reflect an attempt to perpetuate the status and role he developed while trapped.

K12: A Survival Leader

After the bump struck, K12 heard a half-buried miner shouting:

> He was hollering. I don't know whether he was hollering at anything, but he was hollering. I said to him--I was in the dark--I said to him, "Just keep quiet, just stay right where you are, don't try to get out or anything; just wait till I get hold of my lamp and I can see." I reached around and got hold of the cord and tore the light out from under the stones. I shone it down on him and I said, "Now, be careful," you know. I thought he might have a leg caught and break it trying to get away.

Before K12 could act, G12 uncovered the man:

> I just stayed right where I was until--I wasn't sure whether anything was wrong--I stayed there for, I don't know, maybe two or three minutes. One of my first ideas was to get my head down because I knew I was tight to the roof and the air was quite heavy. I imagine there might have been a percentage of gas or something, so that was my first idea. And, of course, G12 went by me to help the other boys out. But then, I don't know just exactly what I done.

Later K12 joined the rest in their escape attempts, but provided little if any leadership. Although he attributed a greater number of "I" initiations to himself for the escape period than did any of his companions, the group credited him with the second lowest rank.

When the group of twelve found their way blocked in a dangerous passage on the third day, with no water, little energy, and very little light, K12 suggested that they get out of that spot and return to the relative safety of the cavity in the 13,000 foot level wall. The men agreed, and they moved back in relays. This was apparently the turning point in K12's role in the group. For the survival period, his Group Evaluation score was twice as high as that of any other man.

After K12, along with N12 and Q12, found a can of water, he took charge of rationing it and sent the men with rations for the two injured men (O12 and P12) who were about 50 feet away. He was at the broken air pipe when he heard the rescuers working and he stayed by it from then on, shouting through the pipe, with others spelling him off. When the rescuers finally heard them, K12 demanded that the rescuers send water in to them through the pipe. He received and distributed the water. A companion said,

> K12, he kept us all from more or less cracking up ...He would say, "Now listen, boys, now listen," and he would explain it out to us. "We can't make our way out and (the rescuers) are coming in to us, now we will just take it calm and cool. It's pretty hard but we will have to do our best." I don't know yet how he done it, but he really done wonders. He kept us pretty well huddled together.

In addition to being instrumental in satisfying two important needs of the group, the needs for water and for hope, K12 handled individual problems in a way to avoid dissension in the group. He felt guilty when the two injured men accused him of drinking water while they slept, and he made sure they got their share thereafter. He forced the weak member, J12, to take water to the two injured men when his turn came because "I knew that if I didn't keep at him to move or something that the other fellows would growl." His own morale was generally good. He had a "rough time" for a day or so, but he said, "I figured that we were going to get out because I was determined that I heard the pounding and that they would be coming."

K12 was 37 years old and had had 17 years in the mines. He stopped school at grade 8 when he was 13. He worked at various jobs until his marriage when he went to work in the mines. He tried contract mining but "figured it was too heavy work for a man of my build (5 feet 9 inches and slim) so I decided I had better... get the next best thing to it." Thus he became a "shiftman," a man with minor foreman status. K12 was in the 1956 explosion but was rescued after six hours; he was slightly gassed when brought up. His wife had more formal education than he and was working as a typist. Discussing his in-law problems and sexual difficulties, he said, "We kind of got our difference ironed out...It's a lot better now." He seemed to feel somewhat inferior to his wife and a little dependent on her. K12 confined his associations to his family and a small circle of close, long-standing friends. He apparently did considerable cooking and enjoyed it, and he built his own summer cottage. He appeared to work things out in an open and practical manner, avoiding conflict and dissension as much as possible.

K12's condition on rescue was fairly good. When interviewed seven days after rescue, he gave the impression of a rather quiet person, slow and deliberate in speech. On psychological tests eight days after rescue, K12 obtained a prorated IQ of 77. This was ten points below the group average, and in view of his high Vocabulary score, ten, it was inferred that his intellectual abilities were quite seriously impaired. On the other hand, practical judgment, planning, and organization were good. His underlying personality was well integrated with considerable resources in the way of control and realistic flexibility. Moreover, his reality testing improved under pressure. Some tension was in evidence, and although emotional expression was being tightly controlled, he did show signs of empathy and awareness of dependency needs. His sentence completions were quite unusual for their candor, considered awareness, and expressive qualities. He was a person who had considerable self-awareness, acknowledging and facing up to problems. In his interview, K12 reported more initiations than any other man for both the escape and survival periods. He apparently was very aware of events and felt free to recount them. Like D6, he attributed a high proportion of initiations to himself.

The two survival leaders, D6 and K12, shared a number of characteristics:

    (a) They were sensitive to the moods, feelings, and needs of others, rationalizing and sympathizing with them when appropriate.

    (b) They sought to avoid conflict and dissension.

(c) They were intellectualizers, using communication rather than action to satisfy the group needs, their verbal abilities being better than their performance abilities.

(d) Their role in the survival period was to a considerable degree a function of their need to have the general good opinion and recognition of the whole group, rather than the specific good opinion of a special friend or partner.

(e) They perceived themselves as making an important contribution to the group.

Q12   A Willing Follower

Q12, a 26-year-old miner with seven years in the mines, gave the most clear, orderly, and detailed account of the entrapment. His role was that of an active, willing worker in all proceedings, an active participant in escape and particularly in survival operations, he appeared to follow very largely the suggestions of the others, especially K12, the survival leader.

> There was a hole there, you could crawl up top... They could smell gas, so they came back...We stayed there a while. We wasn't satisfied. Try it again. So away we went up the wall. This time they decided for me to go first. I got about 15 feet away from them, and I just blacked out.

When they found the last possible escape passage blocked, Q12 supported K12's suggestion that they return to the top of the wall where it was safer and rescuers were more likely to come through first. Q12 accompanied K12 and N12 in a successful exploration for water. During the survival period, he remained with K12 by the broken air pipe through which they heard the rescuers and finally made a contact with them.

When interviewed 18 days after rescue, Q12 was neat, well-groomed, and appeared healthy and fit. He showed no signs of fatigue, tension, or anxiety, and appeared to have few aftereffects, perhaps fewer than any of the other trapped men.

Q12's schooling did not continue beyond grade 7, which he reached at the age of 14. His mother died when he was 11, after which he was raised by his grandmother. His work record was more orderly than many of his fellows two years in a truck as a driver with his father, two years driving a taxi, and two years in a garage. He sought work in the mine to earn more money. He was invited to move up from datal to contract work after three years in the mine. After marrying at the age of 23, he built his own house. He did not drink and smoked moderately. The psychiatrist concluded that this man exhibited considerable initiative, persistence, and stability within the framework of his community culture.

Q12 was given psychological tests 18 days after rescue. He obtained a prorated IQ of 100, showing no obvious impairment of primarily intellectual functions. His performance abilities and practical judgment appear to be his forte, and they were apparently untouched by anxiety. His handling of unstructured material indicated basic stability and better than average resistance to disorganization. Moreover, his reality testing improved as demands on him were increased. No tension was observed, emotional responsiveness tended to be impulsive and spontaneous rather than modulated. He exhibited no signs of empathy or awareness of dependency needs. The Sentence Completion test indicated that Q12 was evasive of emotional involvement and self-criticism, had rather egocentric interests, and had little capacity for independent and imaginative thought. It was inferred that he would rely on others in making important decisions. On the whole, this man appeared to be a "performance type" who relied on intuition and practical abilities, he showed signs of repression or lack of feeling, and his interests were focused on his own wishes and welfare.

This young man's predisaster history reflected better than average achievement drive, persistence, realism, and stability. His behavior during entrapment was in keeping with this picture, and he was in remarkably good condition when interviewed and tested. The interval between rescue and examination was longer for him than for most others, but this alone would hardly account for his apparent freedom from aftereffects, as examplified by his traveling over one thousand miles to see about a job just five days after rescue. A more likely inference is that Q12's ability to remain emotionally uninvolved and his lack of imagination enabled him to shrug off and forget much of the emotional impact of entrapment. Moreover, the situation offered considerable opportunity for the exercise of his intuitive, practical approach to problems. And finally, in the survival period he took the role of lieutenant to his older colleague, K12, from whom he may have drawn support. His recognized contribution to group support in the survival period indicates that his own mood and morale was relatively positive and steady. It is probable that his association with the recognized leader in that period, as the latter rationed the last can of water and heard rescuers through the air pipe, also induced the group to perceive Q12 as contributing to its needs.

J12: A Passive-dependent Man

J12, a 22-year-old datal worker with three years in the mines, was dazed and confused by the bump. He reported, "I was a little shaky and nervous...and didn't quite know where I was " He made escape attempts with the others for a time, "Then I don't know--it just seemed as if we were dozing. Some of the boys went up the wall and dug around for water cans."

J12's story was vague and evasive. Others said that he was at times emotional and helpless

> J12, he wouldn't move at all, and I would threaten him with "Well, I'll give you no water." He would cry for a little. I really shouldn't have done it, but I knew that if I didn't keep at him to move or something, that the other fellows would growl.

His companions gave him no credit for initiations. He himself recalled a higher than average number of initiations, as if well aware of events during entrapment, however, the reliability of some of his statements is in doubt.

On rescue, J12 appeared to be less exhausted than the other trapped miners, and he wanted to go home. He did not want to be alone. When interviewed seven days after rescue, he was somewhat tense and evaded emotional material. He seemed to enjoy lying on a couch, being waited on by his wife, receiving attention, and being something of a hero. He said that he was sleeping well, had no dreams, and had no complaints other than a sore back. He had made little effort to go out since being rescued and displayed no urge to see how the other men were getting along. It appeared that he was giving his strong dependency needs full reign. During the month following rescue, he had several attacks of "trembling and crying."

An only child of parents who separated when he was 14, he said that he got on well with his mother but his father was "contrary." He reached grade 8 when 17, after failing two grades. His wife appeared mentally retarded. They had two children, one of which died in childbirth. While mixing well with a large number of people, he was not a member of any community organization. He liked tinkering with cars and playing country music on a violin.

J12 was tested seven days after rescue. His prorated IQ was 87, his range of ideas was greater than average, and he exhibited high productive energy. His abilities were relatively even at the average level, with no evidence areas of impairment. Judgment and practical performance were virtually intact and apparently unaffected by anxiety. Rorschach control was very good, with less than average disorganization. There was some evidence of tension, but emotional responsiveness was covered up. This man exhibited a considerable wealth of resources and gave the general appearance of being well integrated and stable. However, his resources were not utilized for constructive purposes but tended to be diverted into fantasy. Moreover, when situational demands were increased, his reality testing tended to deteriorate somewhat. An independent clinical appraisal of this man's Rorschach was very positive.

Psychological test findings were at variance with the behavioral picture, indicating a self-contained and well-integrated personality with little impairment of intellectual and performance abilities. The only obvious explanation for this man's behavior under stress is that regression to complete dependency was his habitual means of handling stress, although this was not brought out by the tests. It is notable that this defense apparently protected J12 both physically and psychologically.

F6. A Disorganized Man

F6, a 37-year-old contract miner with 17 years in the mines, was buried to the waist by the bump. E6, the escape leader, came to his aid when he called for help.

> This here hip I thought was broke. I couldn't put no weight on it. We got to the top of the wall and tried to get out but we couldn't get out, so I had to lay there mostly for about two days...I didn't get out, I didn't crawl out. The boys told me it was only about a foot high. Well, I couldn't crawl out myself because my hips were too sore...They left me, the other five went out of the top of the timber way. They took an axe and a saw and were going to try to get through and they left me with the pinned man. He was caught there, and him and I talked.

He denied having any psychophysiological symptoms or hallucinations; he also denied being scared or "bothered one bit." However, there were considerable evasion and vagueness in his story, and a number of inconsistencies. He was obsessed by the thought of drinking pop and talked about this often, much to the annoyance of the other miners.

> I thought about that pop and ice cream. It near drove me crazy, I'm telling you. That was on my mind...most of the time. It was getting on the boys' nerves so I didn't say too much about it.

He said he had no hope and expected to die there

> I thought we were there to stay...I started to give up hope, in fact I did give up. I was prepared to go, and I wasn't one bit scared. I made up my mind. I said, "Boys, we've had it now...We'll all go together!"

He was bothered by the raving of the pinned man and had to be restrained when he impulsively decided to chop off the man's arm. Several of the others said that F6's behavior was poor and that he came nearest to breaking.

On rescue, this man walked most of the way to the ambulance, despite his sore hips. He required an extra day in the hospital. When interviewed six days after rescue, he was untidy and unshaven. He was restless and tense at times. He said that he was sleeping well and "never thinks of it (the experience) now."

After several failures, F6 quit school in grade 4, at 14 years of age. He married when he was 20 and has a son aged 16. He had had bronchitis or asthma for several years and was given psychiatric treatment for acute alcoholism about two years before the disaster. About the 1956 explosion he said, "I couldn't force myself to go down and do rescue work."

On psychological tests ten days after rescue, F6 obtained a prorated IQ of 66. Although there was some evidence of impairment of concentration, there was little to suggest that this man had had average intelligence. He seemed a rigid person with little potential for adaptive behavior, confining himself nearly always to popular stereotypes. On the Rorschach, his elaborations tended to spoil his percepts, suggesting that his reality testing deteriorated under stress. His Bender-Gestalt productions were extremely poor, and although he did them in a hurry, it was inferred that his judgment and practical performance abilities were deficient. Dependency needs were expressed. Signs of empathy were absent, but these may have been repressed. On the other hand, he was preoccupied with fears of the mine and did not show concern about work and the future as many of his fellows did. It is notable that the Rorschach did not reveal impulsivity or tension. It would seem that his self-esteem and mechanisms of denial and repression had been working overtime to present a good front in the ten days after his rescue.

X6. The Semi-isolated Man

X6 was a 42-year-old datal worker and the son of a miner. He was trapped in virtual isolation for 8 1/2 days. He was found by members of the group of six on the second day, but was thought by them to be "pretty near gone." On about the fourth day, however, two miners searching for an escape passage came upon him and were surprised to find he had changed position. X6 spoke a few words to them. Two or three days later, the two miners found he had moved again.

The enclosure to which he had moved was very small, about five feet high and four feet square. There was a hole, however, that went up through the ceiling quite a distance, providing some air circulation, though carbon dioxide may have been trapped in the pocket. If carbon dioxide were present, his prone position would have increased his chance of inhaling a maximum amount of the gas.

The group of six said they did not bring him to their cavity because the way was very narrow and dangerous and they were weak. Thus, X6 was virtually alone until he was rescued.

X6 was seen daily in the hospital. He lay all day immobile in bed with the sheets tucked under his chin as the nurse had left them. He smiled with a warm, rather sickly, childish grin. He replied to simple questions but talked mainly about his head, which was hurting him rather badly, and about the fact that he was not sleeping well. He seemed to tire after two or three questions and would say nothing more. On the fifth day, he volunteered one remark "It is a nice day out today." This was the first spontaneous remark he made to anyone. The nurse who was looking after him felt that for the first few days after his rescue "he wasn't quite with us." This was also the impression of the interviewer.

X6 was interviewed on the twelfth day after rescue, when he was showing considerable improvement. He was seen for one hour, as he sat on the edge of the bed. He seemed dazed and somewhat withdrawn, it took several seconds before he made contact and started to respond. He looked rather sad but smiled warmly on occasion. The reason for the interview was explained, but he showed no emotional reaction apart from a smile. His memory for remote events appeared to be quite good, though his memory for the entrapment was very limited and vague: "It seems as if it was a--a dream." However, he did remember "drinking a bit and eating a sandwich." He also recalled wandering around and trying to get out, "another man hollering at me...standing and falling down again," and he remembered being carried out. As for his thoughts and feelings

> I thought I was gone...I felt pretty bad, I felt half weak, the air was bad and the gas bothered me when I lay down...I thought about my wife and I was worrying about her and what was happening.

When contacted by the other trapped men, X6 apparently was disoriented, for he did not ask where they came from or where they were going, he only asked for a drink of water. He did not drink urine nor, in fact, did he remember urinating at all. He did not remember praying. He did not think he had had a headache, and he recalled no unusual sensory experiences. He remembered what he was doing when the bump struck, but was vague about the three or four days prior to the bump and the first two days in hospital.

X6 did not seem anxious or tense, nor did he tire as much as expected. His major symptoms were memory loss, impaired attention, and confusion. He had difficulty in differentiating time and place, so that at times it was difficult to tell whether he was talking about his time in hospital or in the mine.

X6 described his childhood as being very unhappy. His father drank a lot and "was always fighting" with his mother. He got along well with his mother but not with his father. He appears to have had many nervous traits in childhood.

As for family history, three brothers were in a home for the mentally ill; one was apparently epileptic. X6 described one aunt as "manic." The remaining members of the family were better adjusted mentally. His mother had diabetes and had been blind for some years. He started school at 6 and failed grades 1 and 2: "Things were bad at home and I used to be mad at my father." However, he went on to finish grade 7 at the age of 15. He left school "I wanted to make some money as my father was drinking it all." For ten years after leaving school, X6 worked on a delivery wagon and after that he entered the mines where he has worked for the last 18 years as a laborer. He married when 25, his wife being the only girlfriend he had ever gone out with. Their marriage appeared to be very successful. They had one teen-age son who looked healthy and bright. X6's wife was a large, fat, motherly woman.

His work record at the mine was good. Apart from an accident 4 or 5 years earlier, he had not missed any time. At that time, a fall of stone resulted in a minor head injury, which kept him from work for a few weeks. He described his attitude towards work as "I keep very much to myself." When he drove his car to work in the mornings, he parked it away from the others "in case they would scrape the fender or sides of my car." On the whole, X6 gave the impression of being shy and retiring, a rather schizoid and compulsive person.

X6's outside interests appeared to be few, apart from trying to make some extra money doing odd jobs around the town. He spent most of his spare time cutting grass or hedges in summer or doing rough carpentry for people. "I put up 160 storm windows this fall before the bump." He was paid for all this activity, was very proud of the fact that he did not owe any bills and that he owned his own home and car. He bought nothing on credit. He stayed home most of the time. He visited and played cards rarely. He took no part in community activities and had not gone to church since his marriage.

X6 was given psychological tests 13 days after rescue, while sitting on the edge of the bed. He appeared quite alert, was very anxious to please, and was almost too cooperative and responsive. At times, when he found problems difficult, he would make disparaging remarks about his abilities.

This man achieved a prorated IQ of 112 and his test performance in general indicated virtually no impairment due to anxiety. While his planning and judgment were very good he did not appear to use his intellectual potential to its fullest extent, for he made little effort to think originally. The quality of his Rorschach responses indicated feelings of inadequacy while sentence completion responses suggested one who had a strong achievement drive, was proud, and was rather lacking in self-criticism. There were no signs of tension and he expressed no fear of the mine. There was some evidence of unmodulated responsiveness to emotional impact, and he was quite dependent, especially on his family. On the whole, X6 showed minimum signs of disturbance and better than average resources of intellect and personality.

A subsequent neurological examination (EEG) diagnosed him as having a "low-grade epileptogenic lesion" in the frontotemporal area.

The data on X6 did not provide any clear hypotheses to account for his reactions to isolation in the mine. His extremely spotty memory for the 8 1/2 days suggests that he was in a semicoma and not too conscious of events. This is supported by his apparent inability to do more than flounder aimlessly about and by his evident lack of contact with reality while trapped. His near faultless performance on psychological tests 13 days after rescue also suggested that he had been protected from anxiety in some way. X6 thought he had probably been struck on the head by a stone, but there was no evidence of this except his reported headaches. One possibility is that he suffered a prolonged epileptic seizure produced by subdural pressure from a head blow or precipitated by carbon dioxide (Meduna, 1950). Such a seizure would account for his amnesia, his freedom from anxiety, and his efficient test performance when he recovered. Inspiration of carbon dioxide would also probably contribute to reducing anxiety (cf., Wolpe, 1958) and may have helped to induce a hibernation-like state (cf., Seevers, 1944).

CHAPTER 9

EVALUATION AND SUMMARY

There are a number of methodological problems peculiar to disaster research. That there is seldom a trained observer on the scene at the time of impact raises one set of problems. The research worker has to rely on informants who were in or near the disaster, and he must be seriously concerned with the reliability of information reported in retrospect. The information that informants give is subject to all the errors of perception and recall.

There is a further problem of actual information: Does the respondent actually know? Was he in a position to have the knowledge to be able subsequently to answer the question accurately? It is well known that a person will often answer a question regardless of whether he has direct knowledge of the phenomenon in question.

Within this study a number of people were asked if there were sufficient men available for rescue work. The answers to this question varied considerably. Without doubt, the individuals answered honestly. Whether or not they were in a position to know the answer is beside the point, what is more appropriate is to ask what the respondent perceived in the question. It would seem that to some the question meant, "Did your community respond well to the disaster? Did everyone do as well as he could?" Within this context, they could honestly answer "Yes," saying, in effect, "There were plenty of volunteer rescuers. We did our part." The question might mean to another respondent, "Why did you not perform rescue work?" Within this context, the answer again was "Yes" and meant "There were more than enough volunteers, so I did not have to go." On the other hand, the question seemed to mean to some, "Were there enough men working on your shift, so you could have some respite and be relieved at the proper time?" With this meaning imputed to the question, the answer was "No." The respondent was replying, "On the days that I was working, there were barely enough men to perform the rescue work, and I resented it."

This poses the very difficult question of the use of the nodal informant. Great variations will appear in the information, depending on which nodal informant is interviewed. The question probably elicits more information about the respondent than it does about the problem at hand. This phenomenon is common and widely recognized, but is particularly pertinent in disaster research when the trained observer is not present at the time of the disaster. All answers are "correct" within the appropriate context. What the context is, is another problem. A major task is to identify and categorize the contexts.

The analysis of initiation points to another range of problems. It is quite clear that it makes a difference whether the disaster events are reported by an individual or by all the members of the group. The number of initiations with which the individual credits himself may have no relation to the number of initiations with which the group credits that same individual. As in the case of the nodal informant, both analyses are appropriate, but they are reported from different frames of reference and presumably have different implications. Quantification of initiations does have the merit of cutting through the cover story of the miner.

The limitations of any psychological appraisal are very real, as there may be no way of ascertaining many of the predisaster characteristics of the individual in any reliable way. Much can be inferred from the interview and tests, but there is no way in which the amount of disaster contamination can be estimated. Although a carefully matched control group is of some help, it is only of relative assistance, partly because of incomplete matching and partly because all individuals are involved in some degree or manner in the disaster. A follow-up study may, to some extent, yield information that would provide a check on inferences that have been made.

To study adequately the complex effects of disaster requires the use of techniques from several disciplines. Yet, there are a number of problems in interdisciplinary research resulting from the techniques and assumptions not held in common by the disciplines. The clearest sign of interdisciplinary difficulty in this study was the disappointing attempt to collect many types of data in a single interview. Few interviewers are able to probe to any depth in areas and disciplines in which they have little training or competence. The very thing that interdisciplinary research sets out to achieve--integration--may limit the quality of the research.

From the data made available through these procedures, initiations were the only measure of miners' behavior while trapped. The number of initiations credited to a miner by all the other members of the group was taken as a measure of the extent to which that man was perceived as promoting current group activities. Whether individuals with high Group Evaluation scores should be called leaders is a matter of definition. According to the interactional view (Gibb, 1954), high initiators would be called leaders.

Although quantitative group evaluation of initiation based on the interviews indicated different leaders for the escape and survival periods, the miners' interviews on a qualitative level generally pointed to the survival leaders. Several men spoke in glowing terms of the survival initiators, expressing an emotional and dependent attitude toward them. Such expressions did not occur with respect to escape initiators, although these miners were promoting the group's goals at that time. It appeared that survival initiators were appreciated by the others for their emotional support, while escape initiators did not engender that kind of response.

From another point of view, psychological tests and psychiatric ratings of ego strength favored survival leaders. Indeed psychiatric measures, Vocabulary, and education had negative relationships with escape leadership. They were at variance with the frame of reference of the trapped miners in the escape situation. It is quite obvious that leadership could be cast in a variety of terms.

The fact that the analysis has used initiations as a measure of behavior should in no way imply a judgment that either a high initiation score or a low one is desirable. Rather than evaluate behavior, initiations simply measure and describe it. It remains that an evaluation of the direct or indirect usefulness of initiations is not possible without making a number of assumptions as to what in each circumstance could be considered functional behavior. Many complex factors are involved, and any evaluation of the utility of behavior for either the individual or the group would have to include an ex post facto knowledge of what actually did happen after the bump. A few of the considerations necessary for any analysis of the usefulness of particular behavior can be briefly mentioned.

Immediately after the bump, the miners had two clear alternatives: they could sit, rest, and await rescue, or they could attempt to find a way out of the mine. Attempting to escape had a widely accepted precedent: the majority of the survivors in both the 1956 explosion and the 1958 bump had walked out of the mine. This was a sensible pattern, particularly for miners skilled in the mechanics of mine activities, for it allowed rescuers to concentrate on those miners who were unable to make their own way out.

On a slightly more subtle level, escape attempts provided the trapped miners with activity. Rather than sitting passively to await rescue or death, the miners occupied themselves with activity that in the situation was not random but was organized and appropriate. It would also have been useful if, indeed, they had managed to work their way out through the blocked passages. Such meaningful physical activity would seem very important in light of the fear of insanity or loss of self-control expressed by the miners. Also, attempts to escape, which lead to the conclusion that escape was impossible, may well have been an important precondition to the absence of recrimination and conflict during the survival period.

Two of the three high escape initiators exhibited good psychological test results. On the other hand, test results indicate that some miners who contributed few initiations showed minimum psychological effects. It could be argued that a miner who sits still to conserve his energy while his companions are trying to escape is doing the wisest thing for himself. It could be suggested that this miner saves not only his own resources but also contributes to the group, not in terms of initiations, but by presenting a focus of concern for the group. The inactive man can equally well contribute to the group in terms of stability, reassurance, and calm. From the latter frame of reference, the most inactive man can well be the "leader" if he fulfills other criteria.

If it had turned out that the bump was not as severe as it was and the escape attempts had been successful, leadership might well have been attributed to the most active initiator of the escape period. Under these conditions, a so-called maladjusted person could act in a way that was useful to the group by carrying out his own particular adaptation to the situation. It would be easy for the investigator to overlook the possibility that an unstable personality may be most functional for a group under certain conditions.

On a physical level, however, the unsuccessful escape attempts posed a definite threat. The physical stress, the fatigue, the added danger, and the rapid use of water resources during escape attempts shortened the period for which the group could survive. Even with a post hoc recognition of this threat on the physical level, it is very difficult to estimate how useful prolonged escape attempts were to the group. The majority of the miners seem to have acted on the assumption that they should attempt to find a way out until their lights fail.

A full assessment of the behavior of the miners would involve a study of the motivations, specific consequences that they anticipated, and specific consequences they did not anticipate. The effects of this behavior on each individual, as well as on the group as a whole, would have to be established, but more important, any appraisal would have to be made within a situational context Though it is possible to make evaluations of types of behavior, it would not be valid or useful to do so with only hindsight as to what the situation was. Any evaluation of the behavior of the trapped miners would require the adoption of a frame of reference, and any frame of reference would imply a value judgment.

Summary

The Minetown mine disaster was relatively unusual when contrasted with most other disasters reported in the literature First, 19 miners in two groups were trapped in a coal mine for 6 1/2 to 8 1/2 days and were later available for examination and assessment. The population to which the trapped men belonged was stable, making it possible to examine other groups in a systematic fashion, including a "control" group of nontrapped miners, the wives of both groups of miners, and selected personal service personnel and their wives. The situation thus had some of the features of an experiment in which small samples are treated on a comparative basis. Seventy-six people were interviewed, including 31 trapped and nontrapped miners who were given psychological tests.

The second special feature of this disaster was the long history of previous individual accidents and disasters in this particular mining community, the most recent being a coal dust explosion in 1956. Repeated disasters had provided material for the folklore and expectations of the people in the community. It might be anticipated, then, that individual and group preparedness would play a discernable part in the reaction and behavior consequent to the disaster. It was possible to make a number of observations relevant to this proposition.

When the bump occurred, the shock was felt and heard by many people in the community. Whereas most studies report that only a few people interpret disasters accurately at first, it was quite different in Minetown. Eighty-seven per cent of the 23 subjects having a direct personal involvement, that is, the men having relatives and the women having husbands in the mine at the time, correctly recognized the tremor as a bump. Of the 24 subjects not having such involvement, only 43 per cent immediately identified it as a bump.

The off-shift miners exhibited the convergence behavior noticed in other disasters, all but one of them going to the mine within minutes of the bump. In this case, convergence was due to two general motives: (a) the desire for information about the bump and the welfare of friends and relatives and (b) compliance with the informal miners' code that miners participate in rescue work. It is notable that the code in Minetown was apparently approved by all individuals, including wives as well as miners, for there was no evidence of the conflicting loyalties between family and job duties, which is reported in other disaster studies. While it is true that some of the miners had relatives in the mine and others themselves had been saved in previous disasters, nevertheless, the readiness of the miners to put rescue work before danger and before possible distress to their family was apparently related to the history and folklore of the community. The fact that wives likewise manifested no reservations about their husbands being rescuers indicates that families were also loyal to the code.

The role of preparedness, specifically the experience gained under similar conditions after the 1956 mine explosion, was well illustrated in the behavior of the community during rescue operations. Few of the trapped miners' wives went to the mine, they had been instructed by their husbands to stay away from the pit head because of the crowding that occurred in the 1956 disaster. On the other hand, a great majority of the less involved wives did go to the mine. Rescue operations were organized within minutes. This was largely due to the company's competent disaster organization. Contrary to the findings in other studies (Gordon & Raymond, 1952), no direction from outside the community was required and the entire rescue operation was locally organized and carried out. At the same time, a large number of special services converged upon Minetown from a radius of 150 miles, including Army and Civil Defense units, first aid stations, mobile kitchen, an oxygen unit, and psychiatric unit. It was notable that these special services and personnel integrated rapidly and with very little difficulty. This was presumably because many of them attended the 1956 disaster.

All of the wives with husbands in the mine joined groups of relatives or friends, while only a small proportion of the less involved did so. It was clear that the supportive role of the extended family played an important part in the long waiting period in Minetown. This agrees with the findings of Fritz and Rayner (1958). Presumably the presence of danger and disaster helped to maintain the cohesiveness of extended family units.

There was adequate "official" information throughout the waiting period from radio, television, and press located at the pit head. Nevertheless, there was a tendency to rely on private sources of news, such as the doctor or a relative engaged in rescue work. More than half of the waiting wives rejected the official announcement given on the fifth day that there was little hope for the buried men.

On the whole, individual and group behavior above ground during rescue operations and the long waiting period contrasts with several of the findings in other disaster studies. Minetown's history of accident and disaster had led to the development of a stable social structure designed to handle such events, and the disaster two years previously constituted a rehearsal that made for more functional behavior, especially of the numerous special services that converged on the site.

A content analysis technique was used to identify and count initiations in the miners' recorded interviews. An initiation was defined as an act that originates an extended sequence of behavior. Three types of initiation scores were discussed: Initiation Perception (the number of initiations that a miner reported), "I" Initiation (the number of initiations that he attributed to himself), and Group Evaluation (the number of initiations attributed to him by all the other members of the group). Group Evaluation was taken as a measure of the extent to which each miner was perceived by his companions as promoting current group activities and goals.

An examination of the initiation data revealed that in both groups of trapped miners the interval of entrapment involved two relatively distinct phases: an escape period during which the miners actively sought to find or dig their way out and a survival period during which the miners waited for rescue.

The initiation pattern of the group changed with time and type of problem. The majority of the initiations were reported for the escape period, when the emphasis was on task-oriented activity. Fewer initiations were reported for the longer survival period, when the focus was on affective and interpersonal problems. The miners in the smaller group (group of six) had to contend with the problem of a miner whose arm was pinned near the shoulder. The alternatives were to risk an amateur and probably fatal amputation with an axe or to leave him and hope for rescue. They decided on the latter. The miner died on the fifth day of entrapment. Almost all initiations with reference to this problem were depersonalized, indicating the difficulty of decision on the matter and the strong affect involved.

The two groups differed in their patterns of initiations. There were more miners in the larger group (group of twelve) who played no significant role in the group's activities, as measured by Group Evaluation, thus reflecting an influence of group size on member participation. The group of twelve also mentioned more initiations with indefinite referents, indicating that the activities of larger groups are more depersonalized and a function of subgroup coalitions.

The circumstances and goals of the miners were different in the two periods. In the escape period, they had water and light and their behavior was actively directed against physical barriers. In the survival period, the miners were without water and light and their efforts were concentrated on maintaining emotional control and morale, on satisfying their thirst, and on attracting the attention of rescuers. When initiation scores for the two periods were compared, it was found that the miners' rank order for "I" Initiation and Initiation Perception scores was relatively constant. In contrast, the miners' rank order for Group Evaluation was quite different for the two periods. In other words, the trapped men perceived different members as leaders during the escape and survival periods.

The escape and survival periods were closely analogous to the two types of situations that Bales (1953) has identified. the task-oriented situation calling for instrumental-adaptive leadership behavior and the emotion-oriented circumstance calling for expressive-integrative behavior (cf., Tyhurst, 1957b). An analysis of the interview and postrescue psychological data indicated that the perceived leaders for both groups in the escape and the survival periods had contrasting personal qualities that enabled them to play the required roles. Escape leaders were characterized by the following qualities·

(a) They made direct driving attacks on problems.

(b) They perceived problems as involving physical barriers rather than interpersonal issues.

(c) They associated with one or two friends: the whole group was not their frame of reference.

(d) They were rather individualistic in their expressed opinions and actions, outspoken, and aggressive.

(e) They were not particularly concerned with having the good opinion of most others.

(f) They lacked empathy and emotional control.

(g) Their performance abilities were better than their verbal abilities.

The two perceived leaders in the survival period shared the following characteristics

(a) They were sensitive to the moods, feelings, and needs of others, rationalizing and sympathizing with them when appropriate.

(b) They sought to avoid conflict and dissension.

(c) They were intellectualizers, using communication rather than action to satisfy the group needs, their verbal abilities being better than their performance abilities.

(d) Their role in the survival period was to a considerable degree a function of their need to have the general good opinion and recognition of the whole group, rather than the specific good opinion of a special friend or partner.

(e) They perceived themselves as making an important contribution to the group.

The following psychological tests were given to the trapped miners the Vocabulary, Digit Span, and Block Design subtests of the Wechsler Adult Intelligence Scale, the Bender-Gestalt drawing test, the Rorschach ink blot test, and a sentence completion test. Psychiatrists rated each miner's interview on a number of qualities indicating ego strength. There was a general tendency for test performance to be positively related to initiation behavior as credited to the man by himself or by his mates, however, only the Block Design subtest had significant relationships. Of the other psychological tests, the Bender-Gestalt showed somewhat more consistent relationships with initiations. These two tests, as well as a quantitative Rorschach measure, were efficient in differentiating high and low initiators but did not discriminate their rank order or the middle of the distribution so well. A clinical appraisal of the Rorschach protocols was unsuccessful in distinguishing high and low initiators.

All tests and measures gave better scores to survival leaders than to escape leaders. Some, including education, Vocabulary, and psychiatric ratings of ego strength tended to be negatively related to escape leadership. It was suggested that the frame of reference of psychological tests and clinical judgment is selective and may overlook social qualities that the group recognizes in particular situations. Psychiatric ratings of ego strength seemed to be influenced by the miner's cover story (Dollard, Auld, & White, 1953), giving higher scores to the man who reported more detail and who gave himself most credit.

Psychological tests indicated that being trapped for 6 1/2 to 8 1/2 days with little food and water and with the threat of death affected the miners less than expected. Indeed, their test scores seemed to be affected no more than those of the group of nontrapped miners. On the other hand, the Sentence Completion test indicated that trapped miners were experiencing more subjective anxiety with self-preoccupation, uncertainty, and questioning of personal adequacy. It was suggested that the nontrapped miners did not constitute a real control group, but that both groups had experienced stresses that differed in kind, degree, and the means available to handle them.

There was some evidence that age and years experience in mining were factors in resistance to the stresses involved in the Minetown disaster. Younger trapped miners did better on tests than did older trapped miners, while older nontrapped miners did better than younger nontrapped miners. In the trapped group, age and mining experience were associated with constriction of thought and imagination. It was suggested that this represented an exaggeration of their subjective anxiety to the point of depression.

The group of six miners were trapped for 8 1/2 days, two days longer than the group of twelve. The former group exhibited several unusual trends when contrasted with the latter group less perceptual control and a tendency to overproduce on the Rorschach, more subjective concerns, stronger dependency needs, a tendency to present themselves in a good light, and atypical correlations between tests and initiation measures. It was suggested that a number of factors may have contributed to these results, including length of entrapment, the presence of a pinned miner who died in their presence, and the tighter social controls exhibited in the small group.

The findings summarized in this chapter should not be overgeneralized. They were derived within a situation whose essential features may be relatively unique and they are based on data from rather small incidental samples. However, they concern human behavior amenable to further investigation within social science research as well as within the special field of disaster research. It is hoped that some of the results of this study will be clarified by the follow-up study of Minetown now in progress.

APPENDIX A

THE DISASTER SERVICES IN MINETOWN

Within hours after the occurrence of the bump in Minetown a large number of special service units went into action (see Table 19). Some, like the Red Cross and the militia, were mobilized in Minetown. Others, like one of the army units and the psychiatric team, came from as far as 150 miles. All of these services came to render relief and assistance to rescuers, wives, and the community as a whole. They did not come to help in rescue work. Most of them, including many of the same personnel, had served in Minetown during the 1956 disaster. In general, they functioned as they had in 1956, using the same lines of communication. The following section is a brief summary of the handling of selected community disaster problems.

Traffic Control

During the night of the bump, the traffic to the mine was very heavy. Within an hour, the local police, assisted by local militia, and police from the surrounding region were stationed at important intersections to give the rescuers and equipment the right of way and to direct them to the large parking space in the pit yard. All other vehicles were kept moving to the side streets for parking. The road to the hospital was cleared and kept open so that ambulances had a clear right of way from the pit head to the hospital. At the sound of an ambulance siren all traffic moving along this right of way was stopped by "pointmen" of the three policing agencies.

Rescue equipment, such as oxygen tanks, stretchers, and tents, were moved into the pit yard, some by helicopter, but most of it by truck. Mobile radio and television equipment were also moved into the area.

Organization of Hospital

In the hours after the bump, when the rescued men began to come to the surface in groups of two or three, the entrances to the hospital were not congested and there was ample parking space available at the hospital.

TABLE 19

Surface Disaster Relief Services

Service group	Origin of service group	Time of arrival in Minetown	Location in Minetown	Type of service	Personnel and equipment
Immediate assistance	Minetown	Immediately	Pit head, hospital	Traffic control, transportation, medical aid, canteen	Local police, doctors, nurses, taxis, station wagons
Clergymen	Minetown	Immediately	Non-specific	Spiritual support	6 personnel
Civil Defense	150 miles	Few hours	Armories, pit head	Coordination of services and supplies	15 personnel
Red Cross, Mobile Red Cross	Minetown, 150 miles away	Immediately, few hours	Pit head, church	Groceries to victims' families, clothing for rescue workers	40 personnel, canteen, depot
St. John Ambulance	Minetown, 150 miles away	Few hours	Armories	Emergency hospital, stretcher bearers, emergency morgue	55 personnel, 2 ambulances

TABLE 19 (Continued)

Service group	Origin of service group	Time of arrival in Minetown	Location in Minetown	Type of service	Personnel and equipment
Salvation Army	Minetown	Immediately	Pit head	Hot drinks and sandwiches to relatives of victims	10 personnel, canteen
Womens' auxiliaries	Minetown	Few hours	Non-specific	Assistance to other services	40 personnel
Boy Scouts[a]	Minetown	Next day	Pit head	Errands, services	20 boys
Cadets[a]	Minetown	Few hours	Pit head	Errands, services	20 boys
Legion[a]	Minetown, 15 miles away	Immediately	Legion hall	Hot drinks and sandwiches, traffic control	15 men, kitchen
Militia[b]	Minetown	Immediately	Armories	Traffic control, other services	60 personnel
Police	Minetown, 150 mile area	Immediately to few days	Armories	Traffic control, other services	15 personnel, 3 vehicles

[a] Relieved 12 to 48 hours after bump by Militia and Regular Army.
[b] Relieved 72 hours after bump by Regular Army and Police, to return to normal employment.

TABLE 19 (Continued)

Service group	Origin of service group	Time of arrival in Minetown	Location in Minetown	Type of service	Personnel and equipment
Psychiatric team	150 miles away	10 hours	Armories	24-hour psychiatric service	One to two psychiatrists, nurse
Regular Army	150 miles away	Few hours	Armories, Legion hall	Liaison with civil authorities, traffic control, supplies	75 personnel, trucks, supplies
Liquid air company	Other centers	Two days after bump	Mine, hospital	Oxygen supply on demand	None
Press, radio, and television	Local, national, international	Hours to few days	Pit head	Information	137 personnel with equipment

The three doctors of Minetown were on duty and the 49-bed hospital was prepared for the emergency. The doctors estimated that it would be three hours before the hospital was needed and that 15 to 20 beds might be required. Accordingly, they had 15 to 20 patients moved and preparations completed to receive the injured. As the 81 men emerged from the pit, the company doctor gave them a cursory check. About 20 cases were admitted to the hospital on the first night and 30 more cases were registered as outpatients. The patients were suffering from broken bones, cuts, bruises, and methane intoxication. For the majority, the length of stay was two to three days. When word was received that 12 trapped miners were alive, a field hospital and an emergency operating room were set up in the armories.

Administration of Victims

The first dead bodies were brought up at noon on October 24. All deaths except one were instantaneous and were caused by multiple contusions and crushing. The bodies were brought to the surface at the end of each eight-hour shift. As rescue work was difficult and slow, the number of bodies recovered on each shift was limited to two or three.

The bodies were unofficially identified in the mine by the friends who dug them out, by the location of the bodies in the mine, and by the check number on their lamps. On arrival at the surface before they were taken from the trolley, they were seen by a doctor, an official of the Union, and a representative of the company. All identification was done by this team. If any of these three men was in doubt, next of kin were brought to view the body. Because of mutilation and odor the family was not asked to identify the remains in most cases. Both the identity and the means of identification were marked on a card attached to the body and signed by all three officials. The duplicate of the card was then taken by special messenger to the mine manager's office. From there, the miner's own minister was notified, and he was given half an hour to notify the family that the body had been recovered and identified. If the miner's own minister was not available, a second minister was called. In all cases, the half hour was strictly adhered to; after this, the name was given to the press.

In the mine, a body was wrapped in a blanket and plastic and brought to the surface on a stretcher. After the body had been identified, it was taken by ambulance or station wagon to the temporary morgue, which began to operate at about 3 a.m. on October 24. It operated for two weeks, first in the armories and later in a tent. At the morgue, the body was placed in a metal, airtight coffin liner and sealed. The blanket and plastic wrapping were not removed. The identification card was removed from the body and wired to the coffin liner, and this information was also entered in the morgue book. The body was then sent to the town's single funeral parlor where it was placed in a regular coffin and taken to the family home. Due to the condition of the bodies, nearly all of the coffins were sealed. Five to seven funerals took place each day, usually from the home of the deceased, and the local ministers took their regular part in each service.

Distribution of Information

As the rescued miners came to the surface, before the mine office officially checked them in, most of their families informally learned of their safety through either a member of the family who was waiting at the pit head or friends who passed on word. By the time the miner had turned in his lamp and washed, his family knew he was safe. The more formal procedure of notifying the bereaved was noted earlier.

As the mine office and the manager's residence are only a few hundred yards from the pit head, contact and communication with officials was relatively simple. Immediately after the bump and for the first nine hours afterwards, the company doctor was the only company official at the pit head. All other officials were in the mine. Twelve hours after the bump, the public relations officer and his assistant set up a press release headquarters in the mine office building, where all information was centralized for the 137 newspaper, television, and radio reporters. No official news could be obtained from the pit head, although much unofficial and sensational news was obtained there. The local radio station received permission to stay on the air 24 hours each day to report the recovery of the dead and the rescued. All commercial programs were cancelled and the entire broadcast time devoted to announcements and news concerning the rescue operation, the release of victims' names--with recorded music for continuity. On-the-spot television coverage was extensive, and national and regional networks extended their broadcasting day when appropriate. As a consequence, most people in Minetown remained at home to listen to the radio. The extensive information service was probably instrumental in controlling crowds and reducing the level of anxiety among the population (cf., Conference..., 1953). This together with the fact that there was little to see--no wholesale destruction --resulted in relatively small crowds at the pit head and the morgue.

Psychiatric Service

In the 1956 Minetown disaster, a psychiatric team consisting of a social worker, a psychiatrist, and a nurse joined the other disaster services present in Minetown (Dunsworth, 1958, Weil & Dunsworth, 1958). They offered psychiatric service to all survivors and relatives of victims who requested such service and were available for referrals from other medical authorities, clergy, relief organizations, etc. This group was expanded to four teams which relieved each other in rotation in order to give continuous service. Patients were mostly townspeople. The fact that few of the rescued miners were seen during the disaster period was regretted since during the succeeding twelve months 14 of the 88 survivors were referred for psychiatric assessment.

In 1958, one or two psychiatrists together with a psychiatric nurse were continuously maintained in Minetown from October 24 to November 3. Teams released from regular duties were changed on a rotation basis, each team serving one or two days at a time. Service was available through referral or on a request basis. Only one of the 81 miners rescued within the first 24 hours was referred to the psychiatrist. None of the 19 trapped miners were referred to psychiatry for appraisal or treatment. Word was spread through the town by the local Mental Health Association and by relief agencies that psychiatric services were available, and from October 24 to November 2, 121 individuals, mostly relatives of victims, were seen and treated. Virtually all cases were examined in their homes.

Grief reaction, exhaustion with nervous tension, and acute anxiety reactions were the most common diagnoses. Grief reactions were usually of the hysterical or muted depressive type. Forms of treatment for grief reaction included emotional support and the prescription of tranquillizers and sedatives. Besides serving in their specialty, the psychiatrists functioned in a more purely medical role. The psychiatric nurse played a supportive role with patients and families and accompanied many wives and relatives of victims to the morgue.

Institution of Recommendations

As a result of the experience at Minetown, a Provincial Disaster Committee has been established to provide guidance and assistance in local disasters, thereby putting into effect most of the above recommendations. The Committee includes representatives from Civil Defense, Red Cross, St. John Ambulance, Salvation

Army, and Canadian Legion. Each of these agencies has accepted responsibility for specific services in the event of disaster. In addition, the National Police and the Regular Army supply liaison officers to the Committee. The terms of reference and operating procedures of the Committee have been communicated to all senior municipal officials throughout the Province.

APPENDIX B

MEDICAL FINDINGS ON THE CONDITION OF THE TRAPPED MINERS

Medical reports of local physicians on the condition of the trapped miners after rescue, together with information from the psychiatric interview obtained from 6 to 23 days later, provided factual material that could be tabulated. The information was coded according to the categories and definitions presented in Tables 20 and 21.

At the time of rescue all the miners were dehydrated, had lost weight, and were weak. During the mine upheaval, only 4 miners escaped with minor simple injuries, 12 acquired multiple bruises, cuts, and abrasions, and 3 were severely injured (fractures and severe tissue injuries).

When rescued, 6 men had minor psychological symptoms (minimal signs of anxiety and depression), and 11 were moderately tense--the tension expressed predominately in speech and restlessness. One had marked anxiety symptoms he cried, trembled, and had weak spells lasting beyond the immediate treatment period. The remaining man was confused.

Upon arrival in the hospital in Minetown, all of the 19 rescued miners received homeostatic restorative procedures in the form of intravenous infusions to counteract their dehydration, vitamins by injection, and antibiotic drugs which were administered for the treatment of existing infections and as a preventive measure. Three injured men needed blood transfusions. Two men required immediate major surgical interventions and 5 required immediate minor surgical treatment. Six miners who displayed emotional symptoms were given sedatives and tranquilizing drugs.

Most rescued men recovered after immediate treatment and 10 were well enough to be discharged after two days of hospitalization. Nine were retained in hospital for more than two days. Of these 9, 3 remained in the local hospital for a considerable period of time, 2 for major surgical conditions, and one for rest and

TABLE 20

Distribution by Category of Physical and Emotional
Condition and Treatment of the Trapped Miners

Code[a]	Category of condition or treatment	Number of cases
	General condition on rescue	
g	Fairly good: dehydrated, but otherwise good	2
f	Fair: dehydrated, weak	10
p	Poor: dehydrated, "sick," "shocked," "weak"	7
	Injuries when rescued	
1	Minor simple	4
2	Minor multiple	12
3	Severe	3
	Emotional state when rescued	
1	Minimal emotional symptoms	6
2	Anxiety: tension symptoms moderate	11
3	Anxiety: tension symptoms marked	1
4	Confusion	1
	Immediate treatment in hospital	
N	No treatment	0
H	Homeostatic restorative procedures	19
M	Medical	0
P	Psychiatric (sedatives or tranquillizers)	6
s	Minor surgical	5
S	Major surgical	2
	Condition after immediate treatment	
g	Good	10
f	Fair	7
p	Poor	2
	Duration of hospitalization in Minetown	
2	Two days or less	10
2+	More than two days	9
	Disposal locally after immediate treatment in Minetown	
N	No treatment	15
I	Investigation only	0
M	Medical treatment	1
P	Psychiatric treatment	1
S	Surgical treatment	2
	Treatment at other centers	
N	No treatment	16
I	Investigation only	1
M	Medical	1
P	Psychiatric	0
S	Surgical	1

TABLE 20 (Continued)

Code[a]	Category of condition or treatment	Number of cases
	Emotional condition when interviewed	
N	No symptoms	1
1	Tense	7
2	Anxious, fidgity, restless, tired easily	8
3	Depressed	1
4	Paranoid trends	1
5	Confused	1
	Physical condition when interviewed	
N	No symptoms	12
T-	Symptoms but no treatment required	3
T	Symptoms requiring treatment	4

[a]For application of code, see Table 21.

psychiatric treatment. Of those discharged after two days, one was readmitted with jaundice and treated medically, and one who was retained in hospital for four days was subsequently given psychiatric outpatient treatment.

Three patients were transferred to other centers, one for amputation of a limb following traumatic thrombosis of the femoral artery, one for medical treatment of silicosis and pulmonary tuberculosis, and one for psychiatric investigation not followed by treatment.

All 19 trapped miners were interviewed by a psychiatrist between the sixth and twenty-third day following rescue. Only one of those interviewed did not show any signs of anxiety. Seven were described as tense, 8 were anxious, fidgity, restless, and very easily tired. One of the miners had definite signs of depression, one man expressed paranoid ideas bordering on delusions, and one was confused.

At the time of the interview, 12 miners had no observable physical symptoms nor did they complain about such, 3 mentioned physical symptoms but did not seem to need special medical attention, and 4 still required and were receiving medical, surgical, or psychiatric care.

TABLE 21

Physical and Emotional Condition and Treatment of the Trapped Miners by Code Category
(The code and categories used in this table are explained in Table 20)

Category	G12	H12	I12	J12	K12	L12	M12	N12	O12	P12	Q12	R12	A6	B6	C6	D6	E6	F6	X6
General condition on rescue	f	f	f	f	f	f	f	f	p	p	g	p	f	p	p	g	f	f	p
Injuries when rescued	2	1	2	2	2	2	2	2	3	3	1	3	1	1	2	2	2	2	2
Emotional state when rescued	1	3	1	2	1	1	2	2	2	2	1	2	2	2	2	1	2	2	4
Immediate treatment in local hospital	H	HP	H	HPs	H	Hs	HPs	Hs	HS	HS	H	H	HP	H	H	H	H	HPs	HP
Condition after immediate treatment	g	f	g	f	g	g	f	f	f	f	g	f	g	g	p	g	g	g	p
Duration of hospitalization in Minetown	2	2	2	2	2	2	2	2	2+	2+	2	2+	2	2+	2+	2+	2+	2+	2+
Disposal locally after immediate treatment in Minetown	N	N	N	N	N	N	N	N	S	S	N	N	N	N	N	M	N	N	P
Treatment at other centers	N	I	N	N	N	N	N	N	N	N	N	S	N	N	M	N	N	N	N
Emotional condition when interviewed	2	1	3	2	1	1	2	2	1	1	N	4	2	1	2	1	2	2	5
Physical condition when interviewed	N	T-	N	T-	N	T-	N	N	T	T	N	T	N	N	T	N	N	N	N

APPENDIX C

THE SENTENCE COMPLETION TEST

The incomplete sentences in this test were made up to sample in a systematic manner the miners' verbalized fears, needs, feelings, attitudes, hopes, and expectations with reference to their disaster experience and to their current situation. The instrument was not pretested. While the items were designed for this particular investigation, most of them are sufficiently general to be applicable in other situations, disaster or otherwise. The efficiency of the instrument is discussed briefly in Chapter 7.

Sentence Completion Test

Instructions: Below are 31 partly completed sentences. Read each one and finish it by writing down the first thought that comes to your mind. Do not bother about whether it makes sense to you or not. If several thoughts come to mind at once, write down the first one, and the next ones also. Do not bother about how well you can write; we are interested in your first thoughts about each sentence. Work fast by just writing down whatever thoughts come to your mind as you read each sentence.

1. Home is ....
2. A leader is ....
3. I would never...................
4. The people I work with are ....
5. My ambition ....
6. Religious people ....
7. I would like to ....
8. A friend will ....
9. A man's wife should ....
10. The future is ....
11. What bothers me ....
12. The thing I like about myself ....
13. What I like about church ....
14. The trouble with bosses ....
15. What I need ....
16. Relatives are ....
17. My greatest fear ....
18. What I like about my job ....
19. God is ....
20. The people over me................
21. To move to another place..............
22. When I am afraid ....
23. My greatest worry ....
24. A good education ....
25. I failed ....
26. What I want for my children ....
27. My mind ....
28. My greatest weakness ....
29. When I am old...................
30. My nerves.....................
31. In times of trouble we need ....

REFERENCES CITED

Allport, G. W., & Postman, L. The psychology of rumor. New York: Henry Holt, 1947.

Anastasi, Anne. Differential psychology. (3rd ed.) New York: Macmillan, 1958.

Bales, R. F. Interaction process analysis: a method for study of small groups. Cambridge, Mass.: Addison Wesley, 1950.

Bales, R. F. The equilibrium problem in small groups. In T. Parsons, R. F. Bales, & E. A. Shils (Eds.), Working papers in the theory of action. Glencoe: Free Press, 1953, Pp. 111-161.

Bales, R. F., Stradtbeck, F. L., Mills, F. M., & Roseborough, Mary. Channels of communication in small groups. Amer. sociol. Rev., 1951, 16, 461-468.

Baughman, E. A comparative analysis of Rorschach forms with altered stimulus characteristics. J. proj. Tech., 1954, 18, 151-164.

Bender, Lauretta. A visual motor gestalt test and its clinical use. Amer. Orthopsychiat. Assn., Res. Monogr., 1938, No. 3.

Burnett, A., Beach, H. D., & Sullivan, A. Intelligence in a restricted environment. Unpublished study, Mental Health Division, Provincial Department of Health, St. John's, Newfoundland, 1958.

Caplow. T. Rumors in wars. Social Forum, 1947, 25, 298-302.

Carter, L. Leadership and small group behavior. In M. Sherif & M. C. Wilson (Eds.), Group relations at the crossroads. New York: Harper, 1953.

Carter, L., Haythorn, W., Shriver, Beatrice, & Lanzetta, J. The behavior of leaders and other group members. J. abnorm. soc. Psychol., 1951, 46, 589-595.

Committee on Disaster Studies. The problem of panic. Washington, D. C.: U. S. Government Printing Office, Federal Civil Defense Administration Bull. TB-19-2, June 1955.

Conference on field studies of reactions to disasters. Chicago: National Opinion Res. Center, 1953.

Doane, B. K., Mahatoo, W., Heron, W., & Scott, T. H. Changes in perceptual function after isolation. Canad. J. Psychol., 1959, 13, 200-209.

Dollard, J., Auld, F., Jr., & White, Alice M. Steps in psychotherapy. New York: Macmillan, 1953.

Dominion Bureau of Statistics. Ninth census of Canada 1951. Ottawa: Queen's Printer, 1953.

Dörken, H., Jr. The ink blot test as a brief projective technique. Amer. J. Orthopsychiat., 1950, 20, 828-833.

Dörken, H., Jr. Psychological structure as the governing principle of projective technique: Rorschach theory. Canad. J. Psychol., 1956, 10, 101-106.

Dunsworth, F. A. Psychological findings in the surviving miners. Nova Scotia Med. Bull., 1958, 37, 111-114.

Eaton, J. W., & Weil, R. J. Psychotherapeutic principles in social research: an interdisciplinary study of the Hutterites. Psychiatry, 1951, 14, 439-454.

Frank, L. K. Projective methods for the study of personality. J. Psychol., 1939, 8, 389-413.

Fritz, C. E., & Mathewson, J. H. Convergence behavior in disasters: a problem in social control. Washington, D. C.: National Academy of Sciences - National Research Council, Publication No. 476, Committee on Disaster Studies Rept. No. 9, 1957.

Fritz, C. E., & Rayner, Jeannette F. Some selected observations and case materials on psychological and emotional reactions to disaster. Washington, D. C.: National Academy of Sciences - National Research Council, Committee on Disaster Studies, 1955.

Fritz, C. E., & Rayner, Jeannette F. The therapeutic features of disaster and their effect on family adjustment: some research orientations. Paper read at Conference on Marriage and the Family, Washington, D. C., April, 1958.

Gibb, C. A. The principles and traits of leadership. J. abnorm. soc. Psychol., 1947, 42, 267-284.

Gibb, C. A. The sociometry of leadership in temporary groups. Sociometry, 1951, 13, 226-243.

Gibb, C. A. Leadership. In G. Lindzey (Ed.), Handbook of social psychology. Cambridge, Mass.: Addison Wesley, 1954, Pp. 877-920.

Gordon, A. S., & Raymond, F. Report of mine explosion disaster, December 21, 1951, New Orient Mine, West Frankfort, Illinois. Edgewood, Md.: Army Chem. Center, Chem. Corps Med. Labs., Special Rept. No. 12, 1952.

Halpin, A. W., & Winer, B. J. The leadership behavior of the airplane commander. Columbus: Ohio State Univer. Res. Found., 1952.

Heron, W., Doane, B. K., & Scott, T. H. Visual disturbances after prolonged perceptual isolation. Canad. J. Psychol., 1956, 10, 13-18.

Hertz, Marguerite R. Reliability of the Rorschach ink blot test. J. consult. Psychol., 1934, 18, 461-477.

Horn, D. A correction for the effect of tied ranks on the value of the rank difference correlation method. J. educ. Psychol., 1942, 33, 686-690.

Horsfall, A. B., & Arensberg, C. M. Teamwork and productivity in a shoe factory. Hum. Organization, 1949, 8, 13-25.

Kelly, H. H., & Thibault, J. W. Experimental studies of group problem solving and process. In G. Lindzey (Ed.), Handbook of social psychology. Cambridge, Mass.: Addison Wesley, 1954. Pp. 735-785.

Killian, L. M. The significance of multiple-group membership in disaster. Amer. J. Sociol., 1952, 57, 309-314.

Killian, L. M. *A study of the response to the Houston, Texas, fireworks explosion.* Washington, D. C.: National Academy of Sciences-National Research Council, Publication No. 391, Committee on Disaster Studies Rept. No. 2, 1956.

Klopfer, B., Ainsworth, Mary D., Klopfer, W. G., & Holt, R. R. *Developments in the Rorschach technique.* Vol. 1. *Technique and theory.* New York: World Book Co., 1954.

Larson, O. N. *Rumors in a disaster: observation of rumors and concomitant factors in a disaster situation.* Maxwell AFB, Ala.: Human Resources Res. Inst., Air Res. & Develop. Command, Res. Memo. No. 29, 1954.

Luntz, H. R. *People of Coaltown.* New York: Columbia Univer. Press, 1958.

Malinowski, B. *Science, religion and reality.* New York: Macmillan, 1925.

Marks, E. S., & Fritz, C. E. *Human reactions in disaster situations.* Chicago: National Opinion Res. Center, 1954. 3 vols. (Available from the Armed Services Technical Information Agency as document AD-107 594.)

Mayer-Gross, W., Slater, E., & Roth, M. *Clinical psychiatry.* London: Cassell, 1955.

Meduna, L. J. *Carbon dioxide therapy.* Springfield, Ill.: Charles C. Thomas, 1950.

National Opinion Research Center. *A plan for the study of disasters, prepared for the Medical Division, Army Chemical Corps.* Chicago: The Center, 1950.

Perry, S. E., Silber, E., & Bloch, D. *The child and his family in disaster: a study of the 1953 Vicksburg tornado.* Washington, D. C.: National Academy of Sciences - National Research Council, Publication No. 394, Committee on Disaster Studies Rept. No. 5, 1956.

Prasad, J. The psychology of rumor: a study relating to the great Indian earthquake of 1934. *Brit. J. Psychol.*, 1935-1936, $\underline{26}$, 1-15.

Rapaport, D., Gill, M., & Schafer, R. Diagnostic psychological testing. Chicago: Year Book Publishers, 1946. 2 vols.

Report of the Royal Commission on the bump in No. 2 mine. Halifax: The Provincial Secretary, 1959.

Reicken, H. W., & Homans, G. C. Psychological aspects of social structure. In G. Lindzey (Ed.), Handbook of social psychology. Cambridge, Mass.: Addison Wesley, 1954, Pp. 786-832.

Rorschach, H. Psychodiagnostics: a diagnostic test based on perception. Trans. by P. Lamkan & B. Kronenberg. (4th ed.) New York: Grune & Stratton, 1942.

Scott, T. H., Bexton, W. H., Heron, W., & Doane, B. K. Cognitive effects of perceptual isolation. Canad. J. Psychol., 1959, 13, 193-196.

Seevers, M. H. The narcotic properties of carbon dioxide. NY State J. Med., 1944, 44, 597.

Siegel, S. Nonparametric statistics for the behavioral sciences. New York: McGraw-Hill, 1956.

Solomon, P., Leiderman, P. H., Mendelson, J., & Wexler, D. Sensory deprivation. Amer. J. Psychiat., 1957, 114, 357-363.

Spiegel, J. P. The English flood of 1953. Hum. Organization, 1957, 16 (2), 3-5.

Stokes, III, J., Chapman, W. P., & Smith, L. H. Effects of hypoxia and hypercapnia on perception of thermal cutaneous pain. J. clin. Investig., 1948, 27, 299-304.

Tyhurst, J. S. Problems of leadership: in the disaster situation and in the clinical team. In Symposium on preventive and social psychiatry. Washington, D. C.: Walter Reed Army Institute of Research, 1957. Pp. 329-335. (a)

Tyhurst, J. S. Psychological and social aspects of civilian disaster. Canad. Med. Assn. J., 1957, 76, 385-393. (b)

CPSIA information can be obtained
at www.ICGtesting.com
Printed in the USA
LVHW022323010423
743238LV00004B/67